Beyond the Veil:
Deception, Truth and the Hidden Promise of Science
Further Thoughts on the Effects of 'Conventional Wisdom'

Jeremy Dunning-Davies,

Departments of Mathematics and Physics (retd),

University of Hull,

England

and

Institute for Basic Research,

Palm Harbor,

Florida,

USA.

Richard Lawrence Norman,

Editor in chief, *Mind* magazine

Journal of Unconscious Psychology

ISBN-13:
978-1541117402

ISBN-10:
1541117409

Beyond the Veil: © 2016 Rich Norman,

J. Dunning-Davies

Standing Dead Publications

PO Box 387

O'Brien OR 97534 USA.

Standing Dead—

Let us cast aside the husk, walk away from that which was our name, and leave it as a dead thing. Standing Dead are we—as a tree rent by lightening: now bright and sudden, we who refuse to fall.

Cover design and artwork: Peggi Wolfe (Wolfepaw)

It is an open mind which is the greatest insult to propriety.

Truth is becoming.

R. N.

Contents.

Preface.

Several years have now elapsed since the publication of my book *Exploding a Myth* (Horwood, Chichester, 2007) and, on reflection, it seems little has changed in the world of science. The reasons behind the writing of that book remain, apparently unaddressed by the worldwide scientific community as a whole. Initially, therefore, it might be worthwhile reiterating some of the background to the earlier book.

Having spent several years engaged in teaching and research in a university department, things took an unexpected dramatic turn following a chance meeting at a conference held at Gregynog in North Wales in 1987. It was at this meeting that one of us (JD) first met Bernard Lavenda. Shortly afterwards, we began considering the validity of the so-called Bekenstein-Hawking expression for the entropy of a black hole. Various aspects of this expression caused us concern from a thermodynamical viewpoint. Accordingly we wrote a short letter which appeared without any problem in the journal *Classical and Quantum Gravity* (**5**, L149, 1988). Since it was a letter announcing a new result, we followed it with a full length article which gave more precise details of our argument. This article was rejected but with no adequate reason for such rejection. Since then, our argument has never been even queried. Although not apparent immediately, this incident heralded a beginning of publishing problems for both of us. Over the intervening years Bernard Lavenda and JD have published numerous papers, jointly and separately, on the thermodynamics of black holes but, in all cases, having the articles accepted for publication was rarely straightforward. The same problem occurred in other areas also, such as when we pointed out errors in the original paper by Guth on the theory of inflation. The point raised here is that open scientific discussion was actively prevented by a person, or persons, unknown. It is important to note that it is not a case of one party arrogantly claiming itself to be definitely correct but rather being prevented from expressing an opinion. As David Bohm once said "Science is the search for truth, whether we like it or not". Such a search for truth must include exchange of ideas and subsequent discussion; without that science cannot progress satisfactorily.

During the years between then and now it has become increasingly obvious that the attitude mentioned is not confined to one or two areas of physics but to huge swathes of the subject. Some topics, such as Einstein's theories of relativity, the 'Big Bang' and black holes seem almost sacrosanct and may be critically considered only at the investigator's personal peril. Other areas, such as the work of Ruggero Santilli in Florida and the ideas of the so-called Electric Universe and Plasma Cosmology are seemingly held at arm's length, even though they may – if examined open-mindedly and thoroughly – offer solutions to many outstanding problems facing scientists today. The pernicious effects of so-called 'conventional wisdom' in the areas mentioned were discussed at length in the earlier book. Here it is the intention to re-examine its influence in the light of more modern developments. The approach, though, will be different in that each individual topic raised will be discussed in a totally self-contained chapter. This will mean a degree of repetition of some items throughout the book but, hopefully, will make each chapter a more straightforward read.

As I (JD) have written previously, my personal scientific journey really began under the tutelage of Peter Landsberg who accepted me as a research student at Cardiff University and induced a lifelong interest in thermodynamics in me. However, I was prepared for this journey by two excellent teachers at Barry Boys' Grammar School – Mr Eric Jones (mathematics) and Mr Digby Lloyd (Physics) – and to them I owe an enormous debt of gratitude. Progress was accelerated by my meetings and subsequent friendships with Bernard Lavenda and Ruggero Santilli while, at Hull University, the late George Cole introduced me to astrophysics and cosmology and our chats over coffee produced more than I am certain would have appeared after countless hours slaving over books and/or internet references. Latterly, I have received much support and important collaboration from David Sands, also of Hull University. I would pay tribute also to those other steadfast friends and willing givers of moral support, amongst whom must be mentioned specifically Stephen Crothers, Wallace Thornhill and Donald Scott.

Finally, once again, I wish to thank my wife, Faith, and children, Jonathan and Bryony for all their love and devotion over so many years. I hope my incessant chattering about thermodynamics, relativity, electric universe and so many other topics hasn't bored you too much. I also offer my undying love to you.

Jeremy Dunning-Davies

(24th Sept., 2015)

Beyond the Veil is a series of articles, each of which stands alone as independent and self-explanatory. The 'veil' is all around us, unseen yet omnipresent. It extends through the heralded disciplines of our noble sciences and society, and stretches further within the deepest concealed fabric of the mind itself. The entire of medicine, physics, psychology, social achievement and ethic lie hidden. Perhaps a question will free them. Let us ask.

In a supreme lucky hit, I uncovered the work of Jeremy Dunning-Davies. Upon inspection, each fact stated was proven out. Soon collaboration began with others of like mind which would start the valuable and unfinished search into the connectivities between water, physics and biology. This book represents a few pieces of what we have found within the sciences, history and psychology of the modern man, and regarding the situation in which science and human achievement have evolved. Humanity has gained much, and sacrificed more. This is the fact, which lies *beyond the veil*.

Richard Lawrence Norman

(10th August, 2016)

Acknowledgements.

Firstly, I (JD) would like to acknowledge the help of a number of my ex-students over the years. Specifically among these are Virginia Castellano, Richard Evans and Laura Padget, without whose help it would have been extremely difficult to collect all the material necessary for chapters 12 and 13.

Some of the material included in the following chapters has formed the basis for several previous articles. Specifically, The material in chapters 3 and 14 formed the basis of articles in *Unified Field Mechanics: Natural Science beyond the Veil of Spacetime,* ed. R. Amoroso, World Scientific (to appear); that in chapter 10 was the basis of an article in the Hadronic Journal (vol. **35**); that in chapters 5 and 6 was the basis of articles in *Physics of Reality*, eds. R. Amoroso, L. F. Kauffman and P. Rowlands, World Scientific, 2013. Finally, the material in chapter 15 formed the basis of a piece in *The Individual*, Sept., 2011. However, all except the last were written for scientific readers and appeared in publications not obviously immediately available for a general public. The intention here is to attempt to widen the field of view for these matters which do, in fact, affect that general public.

I (R. N.) wish to acknowledge scientists Arvydas Tamulis, Elio Conte and Paolo Manzelli for their help and encouragement in my autodidactic studies of quantum physics. I would like to thank my friend 'Lupus Alpha' for his kindly assistance in gathering some of the Parkinson's research in the chapter *What Is and What If.* I wish to thank the journals *Quantum Matter*, *NeuroQuantology*, the *Hadronic Journal* and the *World Journal of Neuroscience*, for having the courage to publish my work. I wish to thank Peggi Wolfe for her friendship and superb artwork. I wish to thank my wife for her unflagging support and patience.

1.   Introduction.

Since the publication in 2007 of the book *Exploding a Myth* [1] little seems to have changed in the scientific world concerning the apparently powerful position of 'conventional wisdom' as distinct from a genuine search for scientific truth. It might be remembered that the subtitle of the quoted book was 'Conventional Wisdom' or Scientific Truth? and the hoped-for intention was to stimulate a truly open public debate of this very point. Sadly, no such debate has occurred and the nearest approach to such has been the publishing online of some personally abusive comments re the author of the said book. In the book, after examining the ethical background to various aspects of an edition of the British television programme *Equinox*, attention was confined to an examination of four issues where the influence of 'conventional wisdom' was felt by the author to occur. These four areas were Einstein's Theories of Relativity, the Big Bang Theory, the question of the Schwarzschild Solution to Einstein's Field equations of General Relativity and Black Holes, all areas which have openly flirted with controversy since their very beginnings; and, finally, the less well-known, but, nevertheless, important topic of Hadronic Mechanics. All four illustrate quite graphically the impact 'conventional wisdom' has had and, indeed, is still having on the progress of physical science; the first three retain their powerful positions because of its power and influence, the fourth is heard of very little and, even if its predictions offer the possibility of solutions to existing problems, 'conventional wisdom' decries such should even be openly examined. As also stated in the book, this may be extremely worrying in the area of physical science but, if it occurs in one sphere of science, it will occur to a degree at least in all other areas and that includes medicine. If it does occur in medicine and we will detail evidence you may examine to that effect, then peoples' lives could be at risk. Of course, for the physical scientist, it is easy to discuss this problem and its effects on physical science but less easy to discuss the full range of possible effects in other areas, such as medicine.

Much of the problem alluded to here is illustrated by a swift perusal of the book, *'The Future of Theoretical Physics and Cosmology'*. This book is a collection of articles, the first of which is an introduction which amounts to a general overview of the entire publication, which arose out of a conference held in Cambridge, UK, in honour of the sixtieth birthday of Stephen Hawking. As will be realised immediately, the personnel involved represent a collection of almost all the 'great and good' of the current astrophysical/cosmological hierarchy and so, it seems it could provide an excellent insight into the state of the accepted 'game' and also the mind-set of what might almost be termed the 'opposition players'.

Before considering the actual contents of the various articles in this book, it seems an initial examination into the affiliations of the contributors might yield highly useful insights into some of the reasons fuelling the motivation for their work and, possibly more importantly, their approach to that work. The vast majority of the contributors come from purely theoretical science departments; for example, of the naturally large Cambridge (UK) contingent, all but one come from the Department of Applied Mathematics and Theoretical Physics, the exception being Martin Rees who is associated with the Institute of Astronomy. The other contributors come from a variety of departments, both in Britain and abroad, but most are well-known names on the theoretical side of things. No-one is overtly an experimenter or observer in the sense of Halton Arp for example. Also, it might be remembered that, as far as Cambridge (UK) is concerned, there is a long tradition, going back to Eddington, if not even earlier, of those working research-wise in astronomy and astrophysics originating as undergraduates in the mathematics department. Hence, the background of most, as with so many other supporters of the status quo in astrophysics and cosmology, is totally unlike that of so many advocating the claims of, for example, plasma cosmology and the electric universe. As far as these alternative ideas are concerned, many would probably think of the true beginning being with Kristian Birkeland and his experiments and observations; this an approach continued by such as Hannes Alfvén, Anthony Peratt, and so many others. These brief observations do, we believe, indicate very forcibly one of the problems facing this general field; one group of people, represented very well by the contributors to this book, seems to start from a pure theory and attempt to interpret observations and experimental results in such a way as to support the theoretical starting point, while the second group, represented by those who support plasma cosmology and the electric universe ideas, tends to start from observations and experiments and attempts to build a theory to explain these. In some ways this may be an overly simplistic way of looking at the situation presently facing us but does, we think, give a fairly accurate overall picture and highlights one huge difference between the two approaches.

Of the articles in the book, the first is, as mentioned earlier, an overview of the remainder and is written by two of the editors. This article alone gives an excellent indication of what is to follow. Hence, it begins by considering the first five articles which are grouped together under the general heading of 'Popular Symposium'. Of course, it goes almost without saying that there is massive glorification of both '*A Brief History of Time*' and '*The Universe in a Nutshell*'. On reading them it becomes obvious immediately that there is no way either could be described as a book which furthered the popularisation of science, at least not in any conventional way; although the second possibly does achieve more as far as popularisation is concerned than the first. The first was the subject of a review essay which appeared in the journal, *Public Understanding of Science* (M. Rodgers, 1992, Public Understand, Sci. **1**, 231-234), under the heading '*The Hawking Phenomenon*'. However, the writer probably correctly summed up this book by noting that 'those who finished reading the book doubtless had little difficulty in believing reports that there were many more buyers than readers'. The writer also pointed out that 'a good popular science book should be stretching, but the trouble with this one is that a number of tough concepts which are vital for following the argument are explained at a pace which must bewilder general readers who lack a background in physics'. However, the title of this piece referred simply to the book as a publishing phenomenon which it undoubtedly was but, as was commented on in a subsequent issue of the same journal (J. Dunning-Davies, 1993, Public Understand. Sci. **2**, 85-86), 'the Hawking phenomenon goes far beyond the actual book to the man himself.' It was pointed out how even then, some twenty years before this current book was produced, the power of Hawking's name was such as to ensure that many articles which challenged his work on purely scientific grounds were not successful in finding a place for publication. It seemed, even then, that in some sense his reputation had progressed beyond the purely scientific.

This point was strengthened further on the appearance of the second book referred to earlier, '*The Universe in a Nutshell*' in which Hawking expressed the wish that *his* expression for the entropy of a black hole should be engraved on his gravestone. A suggestion here of a comparison with Boltzmann perhaps? There was absolutely no mention in this book of the fact that the said expression was originally proposed by Bekenstein and, as explained in detail in '*A Brief History of Time*', Hawking initially rejected the expression as incorrect but graciously agreed to it some two years later. This is just one small point that has been overlooked ever since.

Of the other contributors in this section, though, it would seem that only Martin Rees writes in a way clearly accessible to the 'man in the street' although it would seem that this should be the object of the exercise in writing popular science books. It is not without interest, however, to draw attention to the article by Kip Thorne which is devoted to 'Warping Spacetime'. It is of particular interest because there is a picture on the third page of Karl Schwarzschild with a heading explaining that he 'discovered the solution to Einstein's equations which describes a non-spinning black hole'. On the first page is the said solution in the form

$$ds^2 = \left(1 - \frac{2M}{r}\right) dt^2 - \frac{dr^2}{\left(1 - \frac{2M}{r}\right)} - r^2(d\theta^2 + sin^2\theta d\varphi^2),$$

the form which appears in most textbooks. It might be noted that nowhere does Thorne define r and it is not unreasonable to assume that most would take r, θ, φ to represent the usual polar coordinates. Of course, as Stephen Crothers and the present author have pointed out on numerous occasions, this is not the form of Schwarzschild's solution that appears in his original article; in fact, in that article, there is no singularity when $r = 2M$, such as appears here. Since that mathematical singularity might be deemed the 'origin' of the notion of a black hole in general relativity, an obvious problem exists here for the proponents of the status quo but the point is never raised. Hence, as will be noted in a later chapter, there is a genuine query over the real origin of the whole notion of a black hole. However, one thing can be certain, they are, in some way at least, a result of some theory, some mathematical manipulation, but not of observation or experiment.

All the way through the articles in this section, as well as those in subsequent sections, there seems to be conveyed a sense of almost arrogant superiority. This is probably not intentional but it is an abiding impression gained from reading the book that all these people are totally convinced their model is absolutely correct and their position inviolable. This feeling is strengthened enormously by what follows in later chapters. Also, the later chapters reinforce the realisation of the power of mathematics in all this. Mathematics comes across as being central to all. As someone initially trained as a mathematician, I can truly appreciate the power and beauty of mathematics but, early in my days as a research student, I realised that, when you're dealing with a physical problem, you must be able to state the physical meaning of any mathematical result you obtain and this must be in terms of realistic physics, not an explanation developed by drifting off into a land of make believe or science fiction. It is always important to realise as well that any result so obtained is dependent on the actual mathematical model from which you started. I would tentatively suggest that no scientist truly searching for the truth of a situation can afford any degree of smug self-satisfaction – however small.

All areas covered by the various articles are those to which Hawking is said to have made seminal contributions. However, given his physical problems, it is difficult to understand how he has managed to communicate these ideas, if indeed they are all his in origin, to fellow scientists and co-workers effectively. This point is brought home quite forcibly when it is mathematics. It should be mentioned that he apparently communicated via research students. This also came out some years ago in a television documentary in which it was claimed that Hawking communicated with a research student in pictures and the research student translated these pictures into mathematics. Of course, this may be totally true but, given the highly abstract mathematical nature of the material being considered, it might well raise some quite serious questions in people's minds. The abstract mathematical nature could not be more clearly apparent than in the section on space-time singularities. Here singularities are discussed as seemingly almost physically realisable entities before the discussion progresses to happenings in higher dimensions; there is no hint of the possibility that such singularities might just indicate a breakdown of the model, an interpretation that occurs in so many other areas. This sort of discussion is, to my mind, beautiful for a pure mathematician but raises severe problems when it comes to discussing genuine physical reality. Of course, the whole thing becomes more abstract when string theory is introduced and even more so when the discussion of M theory – where the M refers to membrane – begins. However, it might be remembered that the vast body of researchers throughout the world engrossed in studying string theory and, no doubt, M theory, is composed of a wide variety of people with a wide variety of interests. I actually discussed string theory with a young researcher in that field and his view was interesting. He regarded himself as a pure mathematician and viewed string theory from that viewpoint and had absolutely no interest in whether or not his work had any relevance to physics. I confess I regard this as a legitimate standpoint. If, of course, his work turned out at some future date to have a physical relevance, I would regard that as a bonus but, if the mathematics is regarded – as it should be in my view – as a worthwhile intellectual exercise, that relevance should be regarded as no more than an unexpected bonus. Possibly one problem facing researchers nowadays is the constant demand by universities to attract money by way of research grants. This is an understandable stance in experimental disciplines where the cost of equipment and technical support can be enormous, but the situation is definitely not clearly understandable where purely theoretical disciplines are concerned – after all, a mathematician or theoretical physicist may only require a pencil and paper to proceed with some work!

As might be expected, the section on black holes almost takes the existence of such objects as accepted and proceeds to discuss them as possible probes of relativistic gravity, before considering so-called primordial black holes and even black hole pair creation. The section closes with a discussion of black holes at accelerators – an article which could cause no end of problems for non-scientists and is basically a discussion of the ideas which caused so much panic with some people at the switching on of the Large Hadron Collider (LHC). It is interesting to note that this particular article begins by quoting a letter to Hawking from the Director General at Cern in which he talks of researchers at the LHC having witnessed numerous events which are 'consistent with TEV-scale black hole production and, in particular, with extrapolations of your predictions for black hole radiance to higher dimensions'. One wonders if this letter is the source of all those unfounded worries when the LHC was switched on because some people were convinced at the time that it could produce a black hole which might swallow up our world! However, we would contend that most of these ideas are somewhat fanciful and still better confined to the pages of science fiction books rather than true science.

The theme continues in the subsequent sections and one is able to see just from the list of article titles the sort of message being conveyed and that may be summed up by noting that that message is essentially mathematical or, at the very least, mathematically led. Personally, we have the greatest respect for the intellect of those concerned here and also for their combined and individual intellectual achievements but, even a glance at the above list causes us to ask 'Is it really physics?', 'Are these people really close to an explanation of all we see around us and of all that puzzles us about the cosmos?' We have to answer 'No' and, therefore, feel the title of this book totally misleading. The book may represent the future of theoretical physics and cosmology to some, or all, of the contributors but it may be thought by others a somewhat arrogant title in that it ignores so much physical knowledge which could, and should, be relevant to that future. Here one thinks immediately of the lack of reference to anything magnetic or electrical, of anything pertaining to the ideas and experimentally verified facts of plasma cosmology and/or the electric universe. The entire volume is dedicated to a theory based solely around the force of gravity; the much stronger electromagnetic force makes absolutely no contribution in any of this. As people brought up being led to believe the various gravity-based theories held all the answers, we can say in all honesty that finding out about the possible effects of the electromagnetic force in our universe has awakened a completely new outlook on matters. We are now in a position where we can only express amazement that these ideas are not more widely known and accepted. It follows, therefore, that it seems those who continue to advocate gravity-only explanations for cosmological phenomena are adopting a highly blinkered view of things and, in particular, of the vast quantity of physical knowledge backed up by much accurate observation and laboratory experimentation. This then sums up part of the answer to the question posed at the beginning – one of the great strengths of those who believe in the status quo in physical science is an almost unshakeable belief in the absolute truth of the stance he is adopting but that, together with a degree of perceived arrogance, might also be seen as a possible weakness for, once a slight chink is perceived in his armour, the whole edifice could come crumbling down like a house of cards. Before this even begins to come about, however, it is necessary to consider another great strength which is difficult to quantify and, in a sense, identify but is seen through an example in the introduction to one of the above-listed articles.

It is revealed in one of the articles that the author first met Hawking at a conference in Moscow at a time when that person was simply not allowed to travel abroad. However, Hawking did invite the person to Cambridge for a supergravity workshop and evidently his word carried so much weight that that person was allowed to attend the said workshop. In reality, whether the visit was allowed because Hawking's name carried sufficient weight or for some other reason possibly no-one will ever know but it does appear that his name was a factor and that in itself is a manifestation of one of the strengths of this group. Others, who support the ideas of the Big Bang, black holes, dark matter, dark energy and all the other physically strange notions born to support a mathematical framework which is becoming more and more abstruse, have names which are extremely well-known to the general public as of scientists who are at the true forefront of scientific advance. They too could probably have had an influence on whether or not a person was allowed to travel from the USSR for a scientific meeting in those days when such travel was rare. The power of this group is possibly founded, at least in part, on having the ear of people in positions of real power and by having manipulated matters so as to become the scientific darlings of the media. This latter point has meant, on several occasions, that the media in general simply doesn't reply to invitations to attend scientific events which might be deemed anti-establishment.

The above might be seen as painting a fairly bleak picture if any real change is desired in the immediate future. Virtually all the emphasis seems to have been on strengths rather than identifying weaknesses which could be exploited. However, as we have hinted, many of the articles in this book – especially the introductory sections which are often devoted to eulogising Hawking – offer a clue to what may be a major weakness and that is the appearance of absolute belief in their current position and their approach to all the problems. If a mistake, however small, is proved in this position, that would spell disaster and what never seems to concern anyone is that their stance is based purely on a mathematical model to which numerous additions have had to be made already to allow its continued existence. Even though more and more observations are indicating a more prominent role for electromagnetism, this standard model has no place for it. There is talk of plasmas but none is central to the explanations offered for phenomena and neither can it be in the current model. So, what is the suggested way forward?

Those who unthinkingly support the scientific status quo in these matters seemingly occupy a firmly entrenched position and they are going to be extremely difficult to dislodge. The articles in this book make this point very clearly by implication. The problem of how to accommodate electromagnetic and plasma ideas into their framework remains, though, and it is difficult to imagine how this may be achieved successfully. As has been indicated already, more and more results obtained by satellites and probes are indicating the correctness of many aspects of the plasma cosmology/electric universe ideas. For example, some of Birkeland's early results derived from observation and experiment, which were discounted initially in favour of Chapman's mathematical model, have recently proved to be absolutely correct. Needless to say, the true relevance of this has been allowed to pass almost unnoticed but this in itself is undoubtedly a first inroad into the realms of accepted cosmological theories. It seems likely that, if the present trend continues where more and more observations indicate an electromagnetic input into explanations of observed phenomena, more erosion of the present standard position will occur until, eventually, a total reassessment will have to occur. The conventional school cannot continue ad infinitum adding more and more way-out concepts to their model in order to ensure its continued existence. So far, in very recent times, we have been treated to the invention of dark matter, dark energy and even dark flow so that some might suspect the conventional theorists had turned to the 'dark side'. This cannot continue.

One final thought. Several years ago the English comedian Frankie Howard delivered a truly brilliant satirical monologue at the now defunct Establishment Club in London, In it he referred to a well-known theatre critic of the time who was prone to give the impression that he almost thought he was God. Frankie Howard opined 'It's so silly. There are so many of us'. Some might ponder on this!

This volume provides an ideal background to the various issues to be discussed in what follows. All the separate chapters are written to stand alone so that all the material necessary to follow the train of thought is contained in each chapter, including all necessary references. This does mean, however, that allusion to some topics will occur in several chapters. Again, some chapters may contain some mathematical equations, even on some occasions, short mathematical manipulations. However, this mathematical content is brief and may well be ignored; we are confident that what follows may be read with understanding if these mathematical interludes are avoided.

Content summary:

Beyond the Veil is both an extension of *Exploding a Myth*, and also a broadening of the net. Although this work will remain focused upon the impact of restricted thinking within physics, it will also disclose further aspects and details regarding the all too human effects of personality, ambition, paradigm and power as they affect the full proliferation of human potential and endeavor. The veil which separates what is known from a world of hidden fact, the intransigence that stifles the processes which distill truth to advance in its stead, usury, personal and exploitive motives, extends far beyond the limited consequences of degeneration within the physical sciences. The medical sciences are no less affected, and deeper still, the very substance of the social fabric itself. To see and then solve this mystery, may lead humanity toward another, better outcome. We will peer behind the veil of human history, and discover the basic mental topography of modern man and his morality contain within them a basic hidden flaw, a secret which once unbound may lead toward the brightest future. Hope lies hidden. This work is a pathway to that unseen door, and a key.

It is to be noted that the very task at hand remains by its nature, incomplete. It must be so, for what we have found has not been permitted to be brought to honest fruition and evaluation; what is suppressed has been squandered, prevented from reaching its eventual unknown promise. The future of man is held behind an unanswered question. The future is bright but for one thing: an answer. We will bring you the key, and ask.

The second chapter notes that, as the cost of scientific experiments continues to escalate, the moral need to explain abstruse scientific concepts to the public, which ultimately pays for it all, becomes ever more important. Here, by drawing on two examples in modern physics, the need for more genuine openness in this dissemination of information is highlighted and advocated.

In the third chapter it is intended to reconsider briefly some of the objections which have arisen over the years to both the Special and General Theories of Relativity before raising the somewhat provocative question of whether or not either of these two theories is actually required by modern physics. This is one of the chapters which does contain some mathematical illustration but that may be ignored quite safely by all not interested in the mathematics involved.

The next chapter is entitled, *Is This True?* We have uncovered a dread question of concealment, which holds within it the answer to untold suffering. Are we right? We do not know with certainty, but do deeply suspect the answer is affirmative. We believe the case of Royal Raymond Rife stands as testament to the relentless cruelty and indifference of the 'powers that be' toward the health and well-being of the human race, and makes clear the profound detriment and deadly consequence of those powers to the truthful foundations upon which science and progress are based. Has cancer been cured since 1934? Is it possible to treat a great many diseases by way of frequency specific interventions aimed at their pleomorphic basis? If so, the cost of this suppression to humanity is staggering in its proportion. The approach appears sound and replicable. Precious few aspects of this lost work must be re-derived to reap its benefits. If this can be achieved, a cheap, painless cure for cancer and many other diseases will once again be the province of mankind. Let us look and then ask: *Is this true?*

The next two chapters are concerned with thoughts on firstly redshift and modern cosmology and secondly on the so-called 'Big Bang'. Both topics can arouse passions but both really do require further calm, dispassionate re-appraisal. In fact, some time ago, Jayant Narliker and Geoffrey Burbidge raised the question of whether or not the 'Big Bang' is understood. It is not the intention to attempt an answer to that question here but rather, in the cases of both topics, to draw attention to some physics which is often ignored when discussing them.

Next, we peer into *The Enigma that is Light*. Quantum physics operates under a highly accurate theory which is shrouded in a veil of mystery. The insights of several great minds and some analysis may have uncovered the reasons why. It seems that paradox, duality and uncertainty may be descriptive of our human perceptual and deductive limitations alone, and not of physical processes themselves. What forms paradox, reality, or human confusion? Is 'the observer' reducible to a simple concept? *Is uncertainty the deep province of reality and is wave particle duality endemic to physics?* To the last question on both counts we may answer, "No." Specific theory is redefined without paradox. You may enjoy this chapter which attempts to regain reality in quantum physics, while accounting for the most mysterious and spooky of quantum evidence. It appears quantum physics may be functional without duality, paradox and uncertain confusion. We suggest it is the orthodox theoretical *interpretation* which is paradoxical and hence, deserving of scrutiny.

Next, we examine *Some Possible Links Between Drugs and Violence*. This chapter condenses the results of years of research and a-priori analysis to reveal a surprising yet predictable psychical dynamic and conclusion: certain drugs administered to aid mental imbalance may well have a direct link to the rash of violence so peculiar to our age. This topic is entirely taboo within orthodox medical science. It appears that the basic relationships and mechanisms although obvious and clear to see remain 'unrecognized' and are not acknowledged, nor are they investigated. This article spells out the suppressed information and puts the pieces together. Readers may then assess the hypothesis and decide for themselves if a veil of secrecy and profit has allowed the situation to go unabated.

Not too long ago, two announcements concerning astrophysical observations appeared. The first from NASA mentioned the fact that the Voyager 1 spacecraft had detected a 100-fold increase in the intensity of high-energy electrons from elsewhere in the galaxy diffusing into our solar system from outside; the other revealed that a new all-sky map showed the magnetic fields of the Milky Way with the highest precision and proceeded to point out that the origin of galactic magnetic fields remains unknown despite intensive research, although it seemingly assumed that they are constructed via dynamo processes, such as are said to occur in the interiors of the Earth and the Sun. In the next chapter, it is pointed out that plasma cosmology/electric universe theory can, and does, offer viable solutions to these and other supposed problems faced by orthodox cosmology.

Attention is then drawn in the following chapter to a recent article discussing the amount of largely unrecognised structure present in water and which apparently supports experimental work published by Benveniste in the journal *Nature* in 1988. More recent work by Montagnier seems to support both of the above and attention is drawn to this also. A final personal speculation indicates a possible extension of the work. Scientific material in many fields can be affected by the same 'conventional wisdom' and, in this particular instance, in a case which could affect peoples' health.

Next, we will look *Behind the Human Veil*. How deeply does the veil penetrate the human condition? Does it harbor a secret world of tragic limit which implies a future of war, cruelty and endemic unhappiness embedded within the very psyche of man? Yes it does. It need not be so. Here the nexus of this ancient, culturally enshrined error and the bright answer are revealed. Read of the neuroscience, psychology and history which have controlled our race and cast its unhappy lot, and then, look into the basis of human connection. The highest union between man and his world is already within us, hidden, made corrupt and dark, poisoned by design. Hope remains veiled in shadow. Let us look plainly to discover anew: the birthright of man.

A discussion of the controversial topic of nuclear power and the world's energy requirements then follows. Here I (JD) must pay tribute to the late Professor G. H. A. Cole, who pioneered this whole study, and also to two former students, V. Castellano and R. F. Evans, who did so much to help produce this piece of work. Due to well-publicised problems with traditional fuels, it has been obvious for many years that a review of energy sources available has been needed. It has also been required – although not realistically realised – that a good strategy be developed for dealing with all local and global energy requirements. Here attention is restricted to examining some of the claims and problems of using nuclear power to attempt solve this important question.

Closely linked with the last chapter is the discussion of some possible causes of climate change. This is yet another topic consistently in the news over recent years and one where, in many cases, the general public has been left utterly confused. So many controversies have surfaced over such a short period of time and so many statements believed to be true and supposedly supported by solid scientific evidence have been found to be flawed. Here another look will be taken at this topic but, in this instance at least, the possible influence of our Sun will not be ignored. It is not intended to attempt to give answers here but to open up the debate to include some more highly relevant factors which seem to have been conveniently ignored previously. Also, it is hoped this piece will help make more people aware that these other factors could be important and should certainly be considered.

Next follows, as stated in the chapter title, a short digression concerned with the highly abstract notion of negative temperatures. However, this is included to show quite clearly how even seemingly semi-popular scientific magazines, which are readily available to the general public, cannot be trusted always to disseminate true facts in science. This is followed by two chapters which consider firstly the role of mathematics in physics and secondly the issue of the public funding of science. As readers will note, the first topic has been mentioned in earlier chapters but here attention is confined specifically to that subject and it is hoped people will recognise the correct role of mathematics in this general area of physics and note that, in these cases, it must be that the physics is of paramount importance and the mathematics is no more than a tool to help in the work of discovery. This all leads beautifully to the chapter of public funding since it must always be remembered that it is the public – many of whom are not even aware of the details of the research proposed, let alone understand it – which ultimately funds all this work that keeps scientists employed. Here this whole problem will be addressed and people may draw their own conclusions. However, it might also be remembered usefully that, in the words of Upton Sinclair (1935)

"It is difficult to get a man to understand something when his salary depends on his not understanding it".

Is it possible that this thought is at the heart of many scientific problems?

Next we present a statement and a question, *What Is and What If*. This chapter makes a stark and hopeful contrast available to the reader. From the distant and rude insults of a foolish history, the eye ranges over our present intransigence and beyond, toward the answer. Specific therapeutically efficacious compounds are discussed and the exact way they are restricted is made plain, along with the hope they may yet harbor. Particular modern for-profit drug therapies are assessed. The role of, and evidence supporting, physically efficacious informational processes within biology is brought forth, both as it intersects the propagation of DNA and also within the context of potential new therapies which could, if wisely developed, lower drug prices and perhaps create a new non-toxic approach to medicine itself. What lies undeveloped and hidden? We believe no less than the potential health and happiness of mankind. Let us ask: *What If?*

Lastly, we present our conclusions and a few general thoughts. All work written by Jeremy Dunning-Davies unless specified otherwise.

Bibliography.

J. Dunning-Davies; *Exploding a Myth*, Horwood, Chichester, 2007.

G. W. Gibbons, E. P. S. Shellard, S. J. Rankin (eds); *The Future of Theoretical Physics and Cosmology*, C. U. P., Cambridge, 2003.

S. W. Hawking; *A Brief History of Time*, Bantam Press, London, 1988.

S. W. Hawking; *The Universe in a Nutshell*, Bantam Press, London, 2001.

2. Science in the Present Time.

Introduction.

As more and more money is being requested for scientific experiments which are becoming more and more elaborate, it becomes increasingly important to attempt to explain the basic theory behind the work involved to those who, in the end, pay the bill – the members of the general public. Many look on in awe and wonder when told of the Large Hadron Collider. They have little idea what it is or, in reality, what those in charge hope it will do but are carried along on a wave of, quite probably, genuine enthusiasm from those involved. The lack of knowledge, though, is emphasised by the genuine fear felt by some at the belief that, when switched on, this powerful machine would produce a black hole that would swallow up the Earth. Ridiculous as this may sound, there were people who did believe this and were genuinely stressed by the day of the switch-on. The cost of this machine, as well as the enormous cost of running and maintaining it, are almost beyond the comprehension of many members of the general public. Then there is LISA, the Light Interferometer Space Antenna; another project costing vast quantities of money and, yet again, a project funded eventually by an uncomprehending public. The question must be raised as to whether this is an ethically correct position or not. Also, it seems only right and proper for all those paying the bill to be given some idea of the total background position for each and every one of these massive projects. The need for complete openness is emphasised when the plight of so many unfortunate to suffer from a grave lack of food or be in the grip of some presently incurable disease or condition is considered also.

There is little doubt that it would be extremely difficult, if not pointless, to explain the detailed thinking behind some of these modern projects in the general area of cosmology, for example, to the general public. This is not to appear élitist; it is rather that much of the theoretical background is so complex that relatively few professional scientists understand all the ramifications. Hence, how do you explain the background to people unused to the world of the scientist? It is not an easy task but is one that must be attempted and attempted with complete honesty. By honesty is meant the need to explain ALL the background. This would involve making everyone aware if alternative theories and explanations for effects and observations exist. At present, unfortunately, this is definitely not the case.

Discussion of the Basic Problem.

Much of the fear felt by so many as the day of the switch-on for the Large Hadron Collider approached was occasioned by a lack of knowledge of the real situation which arose for at least two reasons. Firstly, the explanations offered were necessarily sketchy because the concepts involved were so complicated and required vast amounts of background knowledge in physics to gain a true understanding. Secondly however, no-one was made aware of the fact that other serious theories abound which made some of the worries pointless.

For over a hundred years now, scientific thought seems to have been held in the vicelike grip of two theories; - relativity and quantum mechanics. However, what of the qualms concerning the theories of relativity and quantum mechanics? It is well documented that many eminent scientists harboured doubts about the validity of relativity – both the special and general theories – from the beginning. Some, such as Herbert Dingle who became deeply troubled by aspects of the so-called twin paradox, formed doubts after initially being passionate advocates of the theory. Unfortunately, once those doubts arose, it seemed that eliminating them became increasingly difficult, if the account of events outlined in his book *Science at the Crossroads*[1] is accurate. Since those early days, little seems to have changed and, seemingly, it is still the case that challenging the validity of the theories of relativity is not a sensible career option. In fact, even showing that the famous tests of general relativity may be explained by other means[2] is regarded by some as a veiled attack on the validity of Einstein's theory. There have been worries expressed also over some points in quantum mechanics almost from the very beginning of the subject. Frequently, these have revolved around the role of the observer and over whether or not quantum mechanics is an objective theory. One man who has considered these points at length is Karl Popper, probably one of the best known philosophers of science. Although he has written on the topics at length, his book *Quantum Theory and the Schism in Physics*[3] proves an excellent source of his views. He expresses the view that the observer, or, as he prefers to call him, the experimentalist, plays exactly the same role in quantum mechanics as he does in classical physics; that is, he is there to test the theory. This, of course, is totally contrary to the so-called Copenhagen Interpretation, which provides the normally accepted position. This alternative view basically claims that "objective reality has evaporated" and "quantum mechanics does not represent particles, but rather our knowledge, our observations, or our consciousness, of particles". As Popper points out, there have been a great many very eminent physicists who, over the years, have switched allegiance from the pro-Copenhagen camp. He cites among these Louis de Broglie and his former pupil Jean-Pierre Vigier, Alfred Landé and, in some ways most importantly, David Bohm. Bohm, himself an acknowledged and deeply respected thinker, wrote a book on quantum theory, which was published in 1951, in which he presented the Copenhagen point of view in minute detail. Later, apparently under Einstein's influence, he arrived at a theory "whose logical consistency proved the falsity of the constantly repeated dogma that the quantum theory is 'complete' in the sense that it must prove incompatible with any more detailed theory". It was this very question of whether or not quantum mechanics is 'complete' which formed the basis of the intellectual struggle between Einstein and Bohr. Einstein said 'No'; Bohr claimed 'Yes'. The whole problem is discussed in great detail by Popper and, for those interested in this important topic, there can be no better reference than the book by Popper mentioned already. It should be noted also that people like Dingle and Bohm who have dared to question what might be termed conventional scientific wisdom have had their position within the scientific community brought into question.

The two enormously expensive undertakings mentioned earlier – the Large Hadron Collider and LISA – have much in common and illustrate well the need for increasing public understanding of some highly abstruse areas of modern science. Worries about the creation of black holes which could swallow the Earth troubled many. LISA will look for gravitational waves emanating from giant black holes. Hence, black holes are mentioned in both projects but what is the public's conception of a black hole and, indeed, of gravitational waves, and how was that conception achieved?

For many years now, black holes have been popular in science fiction and it is probable that, in many cases, the public's perception of what such an object is was derived from some work of science fiction rather than of pure science. This has been augmented by numerous television programmes, purportedly reporting genuine science. In truth, the programmes have reported science but usually only advancing one explanation and ignoring other possibilities. The modern popular conception of a black hole is almost the perfect example of the public being misled as to scientific reality. Although the idea of a stellar body with an escape speed equal to, or greater than, the speed of light goes back to John Michell in 1784[4], the modern notion initially comes from Schwarzschild's solution[5] to the Einstein field equations of general relativity. There are at least two major problems associated with this and both are kept hidden from the public. Firstly, a simple check of Schwarzschild's original article shows immediately that the 'solution' so often quoted and used[6] is *not* actually Schwarzschild's solution. It is a later version due to someone else. The original does not include the mathematical singularity which leads to the idea of a black hole. Secondly, most modern work in this area of physics revolves around advancing explanations which depend on gravity only; the possible effects of any other forces are effectively ignored. However, most of the matter in the Universe is in the form of plasma. As such, electric currents will be circulating and magnetic fields will be playing a role. The electromagnetic force is much stronger than gravity by something of the order of thirty-nine orders of magnitude and there is a school of thought which feels that it is this force which plays the dominant role in the Universe, - not gravity! People advocating this alternative point out that black holes are simply not necessary in their scenario for describing the workings of the Universe. Incidentally, they also note that such esoteric notions as 'dark matter' and 'dark energy' are unnecessary also. However, challenging the popular view is not allowed as it actually raises questions about the absolute validity of relativity and quantum mechanics. This means that the public, which ultimately foots the bill for all scientists do, is not being presented with all the facts before embarking on financing various extremely expensive projects. This is a position which must surely be altered.

Concluding Remarks.

Science should be studied with a totally open mind and any advances should be examined in a like manner. Surely the aim of any scientific investigation is to seek the truth? Probably mankind will always be found wanting intellectually and any solution to a problem will be no more than an approximation to the real truth, but efforts must continue in all areas to find that elusive complete answer. In the meantime, the dissemination of scientific information to the public must be totally honest and open. Where several theories exist, that fact must be openly acknowledged with no thought for protecting vested interest of any sort. The task will be extremely difficult because of the nature of the technical language and theory involved but it must be attempted. If not, the day may come when, disillusioned with science and scientists, the public refuses to continue funding projects

References.

1. H. Dingle, 1972, *Science at the Crossroads,*
   (Martin Brian & O'Keefe, London)

2. B. H. Lavenda, 2005, J. App. Sc. **5**, 299 - 308

3. K. R. Popper, 1982, *Quantum theory and the Schism in Physics*, (Hutchinson, London)

4. J. Michell, 1784, Phil. Trans. R. Soc. **74**, 35

5. K. Schwarzschild, 1916, Sitzungsberichte der Königlich Preussischen Akademie der Wissenschaften zu Berlin, Phys-Math. Klasse, 189

6. see for example:

R. Adler, M. Bazin, M. Schiffer, 1965, *Introduction to General Relativity*, (McGraw-Hill, New York)

3. Does Physics Need Special and General Relativity?

For over a hundred years now, physics and physicists have relied heavily on the theories of special and general relativity, as well as that of quantum mechanics, to investigate problems arising in a wide variety of scientific fields. Originally, as all undergraduates are told, these theories came about in attempts to explain three major scientific problems of the late nineteenth century – the problem of the passage of light through moving media, the advance of the perihelion of the planet Mercury, and the interaction of matter with radiation. However, in all those years since the beginning of the twentieth century when these three theories were born, although the overwhelming majority has succeeded in stifling most of the discussion of the correctness of these theories, small voices have continued to make themselves heard attempting to make more aware of qualms relating to these theories, particularly the theories of special and general relativity. Here the intention is to focus on the, as far as we know, previously unasked question of whether or not the theories of special and general relativity are even required in modern physics. Obviously, any such discussion will necessitate an examination of at least some of the objections to these two seemingly untouchable theories as that is where the basis for raising such an important question arises.

Hence, first some objections to the theory of special relativity will be examined before looking at problems long known to be associated with the general theory of relativity. It will be noted also that some of the results which are associated specifically in many minds with the special theory and are of frequent use in various areas of physics are, in fact, not peculiar to that theory. Then, and only then, the question of the need for these two theories in physics will be raised and suggestions forwarded as to an answer.

The Special Theory of Relativity.

In the nineteenth century, the existence of a material medium, the aether, pervading all space was a generally accepted concept. The supposed mechanical vibrations of this medium were used to explain the wave propagation of light. One great challenge facing experimentalists, therefore, was to detect the actual presence of this medium. At the time, optical experiments were the most accurate available. Easily the best known was that performed by Michelson and Morley in the 1880's. It is well recorded that this experiment failed to detect the physical existence of the aether. In the history of the development of special relativity, this is the first juncture where questions should be raised. Was it actually true that the experiment did fail to detect the physical existence of an aether? The controversy surrounding this seemingly straightforward question continued throughout the twentieth century and is not resolved even today. It is claimed in the vast majority of, if not all, textbooks that no absolute motion was detected but, in truth, the published data revealed a speed of 8km/s. However, this made use of Newtonian theory to calibrate the equipment and was a figure much less than the 30km/s orbital speed of the earth. It was purely due to this second point that the detected speed was less than the orbital speed of the earth that a null result was claimed. It is now claimed by some that modern analysis leads to a different calibration for the equipment and that this, in turn, leads to a speed in excess of 300km/s. The claim is then that the experiment both detected absolute motion and the breakdown of Newtonian theory. This first supposed detection of absolute motion has supposedly been confirmed by other experiments.

However, it quickly became accepted generally that the Michelson - Morley experiment did, in fact, fail to detect the existence of an aether and there then resulted a major challenge to the theoreticians to explain this null result. After much preliminary work by such as Lorentz and Poincaré, Einstein's special theory of relativity emerged as the accepted explanation. However, since those early years of the twentieth century, there has been much discussion of the results of the Michelson-Morley experiment; it being claimed on many occasions that the experiment did not, in fact, produce a null result. The controversy still exists, to the extent that there are plans to perform the experiment yet again in an attempt to establish beyond all doubt the true facts of the situation. Nevertheless, one important piece of physics is invariably omitted from all these considerations. At the time of the original Michelson-Morley experiment and, indeed, at the emergence of the special theory of relativity, the notion of a boundary layer was unknown. Although Stokes had broached the idea in the middle years of the nineteenth century[1], boundary layer theory was introduced only in 1904 by Prandtl. His original publication was in an obscure journal[2] and it was quite some time before the ideas became both known and accepted.

However, if an aether did exist and if the ideas of boundary layer theory are accepted, then the Michelson-Morley experiment, since it was performed on the surface of the earth, would have been performed within the boundary layer between the earth and the aether. At the earth's surface the relative speed of earth and aether would be zero and so, on the basis of this, a null result should have been expected. Ideally, the Michelson-Morley experiment should be repeated, but this time well away from the possible boundary layer. Seemingly this would necessitate performing it well away from the earth and from all other planets. If the results of such an experiment were not null, the existence of an aether could be denied no longer and it would not be mandatory to assume the constancy of the speed of light. An important consequence would be that, as has been shown by Thornhill, the speed of light would be proportional to the square-root of the temperature of the background radiation. In turn, as has been noted elsewhere[3], this would negate the need for the inflationary scenario in the description of the very early universe.

In a series of articles going back to at least 1985, Thornhill has revisited the whole question of the validity of the special theory of relativity. However, he has approached the question from the point of view of a fluid mechanician. More recently[4], he has concerned himself with contrasting the space-real time of Newtonian mechanics, including the aether concept, with the space-imaginary time of relativity involving no aether. By using the theory of characteristics, he showed that the usual Maxwell equations and sound waves in any uniform fluid at rest possess identical wave surfaces in space-time. Also, in the absence of charge and current, Maxwell's equations reduce to the same wave equation which governs such sound waves. This equation is not general and invariant but becomes so when transformed by Galilean transformation to any other reference frame. The same is true of Maxwell's electromagnetic equations which are not general but unique to one frame of reference; in fact, if the argument of Abraham and Becker[5] is followed through to its logical conclusion, it is seen that, in a general frame of reference, Maxwell's equations assume a form which is invariant under Galilean transformation and in which the operator $\partial/\partial t$ is replaced by Euler's total time derivative moving with the fluid, namely

$$^{D}/_{Dt} \equiv {^{\partial}}/_{\partial t} + \boldsymbol{u}.\boldsymbol{\nabla}$$

where $\boldsymbol{u}$ is the constant relative velocity between the two frames in question[6]. The resulting progressive equations are then invariant and apply to electromagnetic waves in a uniform aether moving with constant velocity $\boldsymbol{u}$ relative to the frame of reference. It is what Thornhill regards as the mistake of believing Maxwell's original equations invariant which has led to the Lorentz transformation and special relativity. Also, he would contend that it has led to the misinterpretation of the differential equation for the wave cone through any point as the quadratic differential form of a Riemannian metric in space-imaginary time.

It should be noted that the modified form of the Maxwell electromagnetic equations referred to here has been derived independently on a number of occasions by a variety of people. Possibly most notable among these is Heinrich Hertz, whose derivation of the modified form is included in his 1893 book, *Electric Wave*[7]. This is truly notable because the date precedes relativity by so many years. Phipps[8] has queried whether Maxwell was aware of this work by Hertz and, if he was, why it didn't provoke him to re-examine his equations himself. However, it is possible, even likely, that Maxwell was aware of this work because it is known that he visited America and discussed the possibility of carrying out experiments using an interferometer to check on the possible influence of higher order terms in his theory. It is thought by some that this is what provoked Michelson to set up and perform his now famous experiment. If this speculation is true, the second part of Phipps' query remains as to why Maxwell didn't re-examine his electromagnetic equations. Of course, it is possible that he did but failed to complete a derivation in a moving medium. However, it is probably more important to note that, if Maxwell did know of Hertz's work, then others would have also and it is surprising, therefore, that special relativity came about as it did. Indeed, following Thornhill's reasoning, it may be felt surprising that special relativity, as known today, ever surfaced. The above mentioned paper by Phipps goes some way to explaining this latter query though. He points out that Hertz used a complicated component notation and didn't make use of known vector identities to simplify it. Also, he imposed an unfortunate interpretation on the velocity appearing in the expression for the Euler total time derivative which led to false predictions – for example, the prediction of the creation of a magnetic field by a moving dielectric – which were disproved soon after his death. Hence, Hertz's theory was discarded, but without a true examination of its fundamental mathematical merit. It is easy, and probably correct, to say that this was understandable but, for the future development of science, it was unfortunate to say the least. It is also interesting to note that Phipps points out that observations had been made in the latter half of the nineteenth century which raised queries relating to the familiar form of the Maxwell electromagnetic equations. Why these were ignored, but criticisms of Hertz's ideas were not, is clearly open for future speculation. In this case, however, unlike some others, both Hertz and Maxwell were internationally well-established as scientists and so, the excuse, if proffered, that Hertz (in this case) was not sufficiently well known amongst scientists of the day is simply not valid.

In yet another article[9], Thornhill showed that the equations governing general small amplitude wave motions to first order in the general unsteady flow of any general fluid also reduce to the same wave equation with constant thermodynamic wave speed in the case of a fluid at rest. The said wave equation was shown to hold in a unique frame of reference and is not, therefore, invariant under Galilean transformation. However, it emerged that it will transform under Galilean transformation into a form which is invariant for all other frames of reference. The wave surfaces of Maxwell's equations are then as for sound waves in any uniform fluid at rest. Again it follows that Maxwell's equations will hold only in a unique frame of reference and should not remain invariant when transformed into any other frame of reference. In particular, he showed that the envelope of all wave surfaces passing through any point at any time is, for the wave equation and, therefore, for Maxwell's equations also

$$c^2 dt^2 = dx^2 + dy^2 + dz^2, \tag{1}$$

where c is the constant thermodynamic wave speed. As he pointed out, this is a differential equation and the immediate task should be to solve it; this he does. It is obvious that this equation is

$$ds^2 = c^2 dt^2 - dx^2 - dy^2 - dz^2$$

with $ds = 0$. Thornhill's claim is then that this is where one mistake occurred, and has continued to occur. His contention is that there is no requirement for Maxwell's equations to remain invariant under transformation and that the above expression for ds^2 has meaning in the present context only when $ds = 0$. He suggests that Minkowski erred in apparently failing to recognise that equation (1) above is merely the differential equation of the envelope of the wave surfaces. A further point to be noted at this juncture is that Maxwell's equations, as normally considered, are derived for a medium at rest. It is conceivable that, if those equations had been derived for a moving medium originally, the controversies surrounding special relativity might never have arisen because that particular development might never have been required.

The above situation concerning Maxwell's equations and sound waves then raises the question of whether, or not, mathematics is required to tolerate the same equation being transformed in different ways for different applications. As Thornhill puts it, "does mathematics allow the wave equation to conform to Galilean transformation when it is applied to sound waves, to Lorentz transformation when it is applied to electromagnetic waves, and to either or both of these transformations when it is considered purely as a mathematical equation, or does mathematics insist that the Galilean transformation is unique and must apply equally to all equations so that the same equation must always be transformed by the same Galilean transformation, no matter to what it is applied, or whether it is applied to anything at all?"

It is recognised that the abandonment of special relativity and a return to Newtonian mechanics would result in a backlog of problems requiring conventional solutions. However, the claim is that such problems would lead eventually to the methods of unsteady gas dynamics and the theory of characteristics, such has already occurred in some instances. Thornhill himself has already tackled the problem of the kinetic theory of electromagnetic radiation and derived Planck's formula for the energy distribution in a black body radiation field from the kinetic theory of a gas with Maxwellian statistics[10]. It is in this article that he shows that, if there is an aether, the speed of light is proportional to the square root of the temperature.

In this latter paper, and in a companion one[11], he argues persuasively against another reason for denying the existence of an aether. This asserts that the Maxwell equations indicate that electromagnetic waves are transverse and so, any aether, if it exists, must behave like an elastic solid. Thornhill points out that Maxwell's equations show that the oscillating electric and magnetic fields are transverse to the direction of wave propagation and say nothing about condensational oscillations of any medium in which the waves propagate. The deduction that electromagnetic waves are transverse might be felt an alternative way of claiming the non-existence of an aether. However, if an aether does exist, then, since electric field, magnetic field and motion are mutually perpendicular for plane waves, the deduction from Maxwell's equations would be that the condensational oscillations of the aether are longitudinal, in analogy with sound waves in a fluid.

Further, as has been pointed out by Thornhill[12], the reason Lorentz 'invariance' gives so many correct results is because one consequence of the Prandtl boundary layer theory is that the viscosity of the aether ensures that the local aether moves with all observers and all observers who move with the local aether have the same unique local wave-hyperconoid given by the differential equation

$$(dx/dt)^2 + (dy/dt)^2 + (dz/dt)^2 = c^2. \tag{2}$$

This follows since the general wave-hyperconoid

$$(dx/dt - u)^2 + (dy/dt - v)^2 + (dz/dt - w)^2 = c^2$$

is invariant under Galilean transformation and, locally, $u = v = w = 0$ for all observers in their rest frames. Again, as noted already, the invariance of (2) between all observers is established by using Galilean transformation, Newtonian mechanics and the aether concept.

Hence, it would appear that there are genuine points of concern over the total validity of the special theory of relativity. However, it must not be forgotten that another major accepted consequence of the theory was that energy and mass are related via

$$E = mc^2.$$

However, is this actually true?

One man who, over a period of time has produced much interesting and relevant material is Harold Aspden. Early in his later writing[13], he reveals some very interesting facts which, while probably well-known to some, will, I suspect, be far less well-known to the vast majority. He points out that physics, particularly electrodynamics, made tremendous and very rapid progress in the later years of the nineteenth century. One of the highpoints of this had to be the discovery of the electron by J. J. Thomson in 1897. This, of course, is well-known but what is less well-known is that this was followed, in 1901, by Kaufmann's discovery[14] that the electron's mass increased with speed. In fact, Kaufmann actually measured variation in the charge/mass ratio with increase in speed. The immediately obvious point concerning this piece of information is that it clearly predates Einstein's 1905 paper introducing his special relativity. It is also worth noting, because it is often either forgotten or deliberately ignored, that the explanation for this variation with speed had been provided by Thomson and others before the advent of Einstein's special relativity. Aspden has obviously delved very deeply into the scientific history of the now famous formula linking energy and mass and this is to the benefit of all, whether or not individuals agree with his conclusions. He notes that, as far as the formula $E = mc^2$ is concerned, definite reference was implied in a book of 1904, - *The Recent Development of Physical Science* by W. C. D. Whetham - where there was reference also to a suggestion made by Jeans to the effect that the energy of radioactive atoms might be "supplied by the actual destruction of matter". In other words, in an article of 1904 published in *Nature* (vol.**70**, page 101), Jeans directed everyone's attention to the store of energy which was available by the annihilation of matter, "by positively and negatively charged protons and electrons falling into and annihilating one another, thus setting free the whole of their intrinsic radiation". Jeans further noted that, initially, he felt he was advocating something new but actually found that Newton had anticipated something similar two centuries earlier, as is recorded in Query 30 of the 1704 edition of *Optics*. However, returning to the question of the equation $E = mc^2$, as Aspden notes, while specific reference to it does not appear in Whetham's book, all the necessary background physics is well presented in mathematical terms. No doubt, Thomson had arrived at his result by assuming the energy of the magnetic field due to the motion of a charge e at a speed v to be $e^2v^2/3ac^2$ and thinking of this as equalling the kinetic energy $mv^2/2$. The equality of these two expressions results in:

$$mc^2 = 2e^2/3a,$$

where the expression on the right-hand side is the energy Thomson recognised as that of an electron with its charge contained within a sphere of radius a. Hence the implied equivalence of mass and energy is deduced.

Again, it should be noted that J. J. Thomson himself referred to the relation $E = mc^2$ in a series of lectures he delivered at Yale University in 1903. These lectures also appeared as a book[15] published initially in 1904. Hence, it is undoubtedly the case that this most famous of physics' relationships was both known and used well before the advent of Einstein's special theory of relativity. Indeed, more recently, J. P. Wesley[16] has noted that this relation is an experimentally verifiable fact and has shown that, by accepting that, he has been able to deduce other relations normally accepted as being linked solely with the special theory of relativity. Possibly the most important example is the following:

Wesley diverges from traditional Newtonian mechanics as a result of his noting that, since mass/energy equivalence is an established fact, if this applies to any form of energy, it follows that there must be a mass equivalent for kinetic energy. This fact has to be included, therefore, in traditional Newtonian mechanics as a modification but, when so added, leads immediately to

$$T = mc^2(\gamma - 1)$$

where T represents the kinetic energy. Of course, this is a result normally associated with the Special Theory of Relativity.

This is achieved without recourse to the traditional notions of special relativity and without use having to be made of the almost totally mathematical approach involving use of the Lorentz transformation equations. This is important because many of the puzzles associated with traditional special relativity may be traced back to the Lorentz transformation equations and it is these puzzles, many of which are really mathematical in fundamental nature rather than physical, which caused Herbert Dingle so much trouble. After many years promoting special relativity, Dingle raised several worries and objections; most notably possibly that concerning the seeming non-symmetry of the problem of the so-called 'clock' or 'twin paradox'. Whatever a person's personal views may be, it is undoubtedly true that the history of this dispute (fully documented in the given reference[17]) hardly indicates a satisfactory resolution of a genuine problem. Here, after all, was a major query being raised by one who had been a very genuine supporter of the special theory of relativity as put forward by Einstein and, once again, a person well-known and well-established in academic circles. Dingle experienced real concerns over the validity of the theory and, as well as those, he recognised that there were in existence *two* special theories of relativity, one attributable to Lorentz and the other to Einstein. The difference between the two, as he pointed out, was a big one; the first retained the concept of an aether, the second did not. However, possibly the most worrying aspect of the case of Dingle is the attitude of fellow scientists to his persistent querying. All recognise that, if someone continues returning to the same old question regardless of the reply given already in hopeful answer, patience can wear a little thin but, when one has read and digested Dingle's book[17], the conclusion has to be reached that a full, frank and totally open discussion of the points raised did not occur but, if it had, it would have been in everyone's interest.

The General Theory of Relativity.

As far as the special theory is concerned, it is undoubtedly true that controversy has simmered just beneath the surface from the very early days. The general theory, however, seemed to offer the only solution to problems which had been taxing theoreticians for some considerable time. Doubts were expressed but, as has so often been the case where Einstein's theories of relativity are concerned, the doubters were regularly dismissed as mere cranks. Again, though, as in the case of special relativity, not all the facts are made readily available to modern day audiences. In Newtonian mechanics, although not specifically mentioned usually, the effects of gravity are assumed to propagate at infinite speed. This follows from Newton's original concept of action-at-a-distance. More recently, the thought has developed that, in reality, gravitation propagates at the speed of light. One example that originally caused problems was the value of the observed advance of the perihelion of the planet Mercury. Newton's theory explains an advance of the perihelion but not of the observed magnitude. It is proclaimed nowadays that Einstein's general theory of relativity was the first to explain the advance correctly. It is true that it does predict the correct value for the advance but, as Aspden[13] reveals, Einstein wasn't the first to offer a satisfactory explanation. This honour falls to a German schoolteacher, Paul Gerber, who presented a theoretical argument giving the precise value of the anomalous advance of the perihelion of Mercury in an article entitled *The Space and Time Propagation of Gravitation* and published in 1898[18]. Gerber actually derived exactly the same formula for the advance as that given by Einstein in

1916 and, in fact, had assumed that the effects of gravity propagated with the speed of light, in common with ideas of today. Aspden comments that Gerber may have made mistakes in his argument but implies that the basic argument was correct and all that was needed was for someone to tidy it up. Instead, this work was, and still is, virtually unknown. This is surprising because the article addressed a major problem of the time and the fact that it appeared in German would have posed less of a problem to international audiences then than it might now.

The arguments surrounding the advance of the perihelion of Mercury and other phenomena supposedly explained by the general theory of relativity and only by that theory have continued apace ever since the theory first saw the light of day. Most suggested alternative explanations have been dismissed, often with a sad shake of the head as if to suggest some degree of sympathy for someone who could be so deluded as to think they could even contemplate offering an alternative. Nevertheless, in more recent years, alternative ways of explaining the shift of the perihelion of Mercury and the bending of light rays have emerged. One of the most recent is that due to Lavenda[19]. He set out to explain the time delay in radar echoes from planets, the bending of light rays, and the shift of the perihelion of Mercury via Fermat's principle and the phase of Bessel functions. It is undoubtedly true that he has succeeded in explaining these three phenomena by this means. However, he has met fierce opposition when it comes to publishing this work. Why? Nowhere does he claim to be attempting to usurp the position of general relativity; he merely wishes to point out that some results, at least, may be obtained by means other than use of the general theory of relativity. As he himself says, "Sometimes new insight can be gained by looking at old results from a new perspective." This highly perceptive suggestion by Lavenda might usefully be noted by all who oppose the publication of anything that even appears to question either special or general relativity, or indeed any who oppose publication of anything purely because it fails to conform to some arbitrary element of 'conventional wisdom'. The alternative suggests an amazingly blinkered view, often by some of the publicly acknowledged giants of the scientific world. The only way forward in any pursuit of knowledge is to admit all possibilities. Once you close one door, you instantaneously rule out one avenue of approach and, therefore, possible advance. Intellectual giant though Newton undoubtedly was, everyone is quite happy to query details of his theories, and rightly so. Hence, why is questioning of Einstein's theories regarded by so many as totally unacceptable? From what one reads of the man, that is not a reflection of the position he might have been expected to espouse himself. Also, it is interesting to note that the same attitude does not seem to affect Newtonian mechanics. Of course, Newtonian mechanics is now extremely well-established and is the theory which dominates everything mechanical seen by the majority of people. It is eminently successful. However, no-one seems to have been offended by the analytical approach to the subject as advocated by Lagrange and Hamilton; no-one seems to have been offended by the 'forceless' mechanics suggested by Hertz as expounded in his book *The Principles of Mechanics*[20]. Why then are so many so apparently over-protective of Einstein's theories of relativity? This is a question to which no-one probably knows the true answer. Nevertheless it is a question which needs to be raised and one of which the public at large should be aware. To emphasise a point raised above, alternative approaches do exist which lead to the solution of problems which may also be solved using the methods of general relativity and, as Lavenda has said, examining these alternatives could lead to new insights.

Mathematics and Physics.

For mathematicians, the general theory of relativity is regarded as a thing of real beauty. This is a position which any non-mathematician may find extremely difficult to comprehend but it is, nevertheless, very true. It must always be remembered that mathematics is a subject which may be studied on at least two very different, but equally important, levels. It may be studied as a purely academic subject in its own right. In this approach, the mathematics is all important and, to the practitioner, can be, and often is, extremely beautiful. It must be noted also that, academically, this approach to mathematics is fully justified; it is a highly worthwhile academic pursuit. However, the second major view of mathematics is as the language of physics. In this context, mathematics may still be seen as extremely beautiful but here it is, and indeed must be, subservient to the physics in importance. Once mathematics is used as the language of physics, it is being used as a tool in an attempt to describe physical situations. It is no longer truly important in its own right. Now, it is the physics of the situation under consideration which is all important and must provide the driving force for any work which ensues. Again, the mathematics is being used in this case to help model a physical situation and it must be remembered always that that is all that is being attempted – to produce a model of a physical situation. It is highly unlikely that any such model will be an exact representation of physical reality; it will be merely an approximation. How good that approximation proves to be is determined by what follows from the theory. Does it, for example, make valid predictions about the physical situation which originally occasioned the investigation? If it does, the accuracy of these predictions will prove a useful guide to the worth of the theory. However, where great care must be taken is in ensuring that the physical situation under consideration isn't, in any way, forced to 'fit' this theory; it is vital to avoid the accusation that observations are interpreted with the predictions of the theory in mind.

The general theory of relativity is one of those topics which rely heavily on very beautiful mathematics, to the extent that the physics of the situation can even tend to be obscured by that very mathematical beauty. Mathematics is a beautiful, rewarding subject in its own right and, academically, no justification is needed to support its study. However, as mentioned above, where study of physics is concerned, mathematics is simply a tool to be used by the physicist in aiding the resolution of a physical problem. In these circumstances, it is the physics which is all important. A theory cannot be adopted to the exclusion of all others simply because the mathematics is beautiful. As far as general relativity is concerned, as has been stated on several occasions, the only results which can be truly trusted are those with a Newtonian analogy. It must be remembered also that, in practice, the results of the theory are used only rarely where descriptions of the physical world are involved; the results are used far more frequently to speculate about the physical world, especially its origins. One must wonder about the worth of speculating about the physical world and its origins on the basis of a purely abstract mathematical theory – however beautiful the mathematics may be. Some of these speculations, which dominate much present day thinking, involve the imposition of a physical meaning to a mathematical singularity. Both the notions of the 'Big Bang' and of relativistic black holes fall into this category.

Conclusions and Final Thoughts.

Much of the above should raise serious questions in the thoughts of any truly open-minded scientists; - indeed, one might sensibly question if someone is truly a scientist if he is not open-minded. As mentioned earlier, Wesley showed some years ago that there is no need to call on the ideas of special relativity to derive some of the more useful equations of physics which are usually assumed to be dependent on that theory. Hence, it seems science is left with only the awkward consequences of utilising the Lorentz transformation equations to explain. In truth, it is these unusual results, such as the 'twin paradox' alluded to earlier and which caused Dingle so many personal problems, which remain but now appear even more isolated from actual physics. Bearing in mind that the need for the General Theory to explain various physical phenomena has been seen unnecessary as well, science is left to consider the important question:

Are the Special and General Theories of Relativity necessary any longer in Physics?

This may seem an almost heretical question to raise but, in the context of present day scientific knowledge, it is certainly one which should be contemplated. Also, just as use of the Lorentz transformation has led to an apparent need to provide physical explanations for what are essentially mathematical results, so general relativity has led to the need for some to demand physical explanations again for purely mathematical results. This has led to numerous publications using up many valuable journal pages discussing what some might term pseudo-problems and also to the ostracisation of those who have queried the validity of the work. In the case of general relativity the relevant examples have been far more serious than the purely theoretical problems arising in the popularly accepted theory of special relativity. It is the basic theory associated with general relativity which has led to both the Big Bang theory and the whole notion of black holes and there can be little doubt that the widespread acceptance of both of these ideas, as well as several others, has resulted in a tunnel visioned approach to the study of many problems in a variety of fields. It should be noted that virtually all the problems to which reference is made, or implied, arise because of a total belief in a theoretical model and an overwhelming reliance on mathematics. As Brillouin[21] has pointed out, 'the scientist should never confuse the actual outside world with his *self-invented physical world model*' and it might be felt that those engaged in research in the physical sciences would benefit from reading Hadamard's book, *The Mathematician's Mind*[22] in order to appreciate the different approaches to research by mathematicians and physical scientists. In fact, the dangers inherent with too much reliance on theoretical models and mathematics has been emphasised both by Brillouin and Rizzi[23], amongst others, and their words of wisdom should be noted and remembered by all who wish to practise true science.

References.

1. G. G. Stokes, 1845, *Phil. Mag.* **XXVII**, 9

2. L. Prandtl, 1904, Proc. 3rd. Internat. Math. Congr.

3. G. H. A. Cole & J. Dunning-Davies, 2001,
in *Recent Advances in Relativity Theory*, vol. 2 (eds. M. C. Duffy & M. Wegener), 51

4. C. K. Thornhill, 1996, Hadronic J. Suppl. **11**, 209

5. M. Abraham & R. Becker, 1932, *The Classical Theory of Electricity and Magnetism* (Blackie & Son Ltd., London) pp. 141-2

6. J. Dunning-Davies, 2002, Hadronic J. **25**, 251

7. H. Hertz, 1893, *Electric Waves*, (Macmillan, London)

8. T.E.Phipps, 2002, Galilean Electrodynamics, **13**, 63

9. C. K. Thornhill, 1993, Proc. R. Soc. (London) **442**, 495

10. C. K. Thornhill, 1985, Speculations Sci. Tecnol. **8**, 263

11. C. K. Thornhill, 1985, Speculations Sci. Technol. **8**, 273

12. C. K. Thornhill, 2004, Hadronic J. **27**, 499

13. H. Aspden, 2005, Physics without Einstein; A Centenary Review, (see www.aspden.org)

14. Kaufmann, 1901, Gottingen Nach. **2**, 143

15. J. J. Thomson, 1904, *Electricity and Matter*, (Archibald Constable & Co. Ltd., Westminster)

16. J. P. Wesley, 2002, *Selected Topics in Scientific Physics*, (Benjamin Wesley, Blumberg)

J. Dunning-Davies, 2013, Hadronic J. **36**, 1

http://viXra.org/abs/1304.0152

17. H. Dingle, 1972, *Science at the Crossroads*, (Martin Brian & O'Keeffe, London)

18. P. Gerber, 1898, Zeitschrift f Math, u Phys., **43**, 93

19. B. H. Lavenda, 2005, Journal of Applied Sciences, **5**(2), 299

20. H. Hertz, 1956, *The Principles of Mechanics*, (Dover, New York)

21. L Brillouin;1964, *Scientific Uncertainty and Information*, (Academic Press, New York)

22. J. Hadamard, 1945, *The Mathematician's Mind*, (Princeton Univ. Press, Princeton, New Jersey)

23. A. Rizzi; 2004, *The Science before Science*, (Press of the Institute for Advanced Physics)

4. Is This True?

Richard Lawrence Norman

and

Jeremy Dunning-Davies

This is the story of a true humanitarian genius named Royal Rife and the facts of reputation and money as they affect scientific honesty and human hope. Science is touted as an objective discipline of fact and deduction but it is an endeavour carried out by people, so it is no surprise that it shares our flaws. "Truth" in science is influenced by the humans who derive it, the social structure in which they work, their personal monetary ambitions and of course those large corporate interests which are so deeply involved have no small effect on scientific objectivity. The paradigms under which the search for truth takes place define what is or is not observed and believed, that is to say, how data is interpreted. Those paradigms are unfortunately, influenced, supported and refuted as a matter of all too human, highly subjective motivations.

Reputation has great bearing upon what becomes known as scientific truth, and the very paradigm under which science is conducted may well be a human question, a question of reputation, corporate and social influence, money (quite naturally) and so *not one* of fact. The result is that the following claims/facts have been brutally suppressed: "Cancer has been cured since 1934. The cure is painless, and cheap. This same method may cure a great many diseases." Could these claims possibly be true? Are they actual facts?

How could such a fact be hidden, if true? Reputation, power, and greed are the familiar answer. These have affected the paradigm under which data is understood, and influenced what has been suppressed or accepted as proper science. Doctors and researchers of high character unknowingly labour under a false paradigm. The cog often fails to realize the machine of which it is a part. To understand the tragic story of Royal Rife, and the even deeper resultant tragedy of the millions of needless painful deaths from cancer and other diseases, we must first examine the history.

Reputation is almost akin to truth itself in science. Science is no different from any other human enterprise, and it is affected by human frailty and hubris. There is a paradigm known as pleomorphism which has been rejected. Disease processes are sustained by way of transformations in biological structures such as cell types which are, therefore, themselves *processes.* Pleomorphism is defined as: The assumption of various distinct forms by a single organism or species. (Dorland's *Illustrated Medical Dictionary*). However, orthodox theory is monomorphic, and does not acknowledge this long observed notion of transformative biological processes.

Many deduced pleomorphism to be valid long ago. Pierre Bechamp was one of them. Louis Pasteur has staked his reputation on the converse view. Bechamp deduced after years of detailed study, that bacteria could change form. Rod like structures, for instance, could become spheroidal but, even further, he noted that the size of these organisms could also vary and devolve into smaller organisms, which were unseen, that he called *microzymas*. This point is crucial. However, Pasteur's reputation was great and Bechamp, whose work was later proven correct, was soundly crushed and his ideas excluded from accepted practice. The paradigm science laboured under for much of the 20th century was thereby hobbled. Thomas Rivers of the Rockefeller Institute derived technical scientific distinction regarding the reproduction of a virus which, although false, cemented his lauded place in the discipline of virology. He introduced the notion that a virus requires *a natural cell* in which to reproduce. His aggressive personality and great monetary resources made him impossible to disagree with, although he was wrong. Dr. Arthur Kendall was unable to defeat Rivers's powerful reputation and formidable personality, but did prove himself scientifically correct by culturing virus strains in an artificial "K Medium" of his own design, and he provided assistance to Royal Rife. Rife, would demonstrate the correctness of the rejected pleomorphic paradigm, and prove over and over that *filter-passing organisms*, meaning very tiny pathogens which are able to pass through filters and may cause full blown disease such as cancer, could be derived from cancer tumours. Rife, would soon discover the impossibly small "invisible" cause of cancer, and allow its direct observation *in a living state* with a new type of microscope which is still unequalled today, and also uncover the lethal frequencies to apply using a specific new instrument to devitalize the tiny bug, which kills man. For this, he would be personally ruined and his work suppressed, at unimaginable human cost.

The cure for cancer and its suppression are an American tragedy which is ongoing. There is a pattern which can be observed repeatedly. Once Kendall and Rife filtered and saw the tiny pleomorphic form of typhus, and proved that this could be done, Milbank Johnson, a prominent and influential physician, USC professor of physiology and clinical medicine, and, Chairman of the Special Research Committee arranged a gathering of thirty top scientists on November 20th 1931 to announce the fact, as was reported by the Los Angeles Times on November 22nd 1931. On December 27th of 1931, The Los Angeles Times reported that Rife had demonstrated his microscope to over 250 scientists. Soon Kendall was asked to speak before the Association of American Physicians on May 3rd and 4th 1932. It was here that Rivers and his cohort Zinsser, whose *Bacteriology* is still used in an updated version today, struck. The attack was verbose and unscientific yet effective in discrediting the correct work of Kendall and Rife, which contradicted their own views. Dr. William Welch an esteemed authority in bacteriology rose to Kendall's defence stating "Kendall's observation marks a distinct advance in medicine," and in August of 1932

Science magazine itself published a report on Rife's microscope stating, "There can be no question of the filterable turquoise blue bodies described by Kendall. . . . examination under the Rife microscope of specimens…leaves no doubt of the accurate visualization…" but the damage was done. Rivers and Zinsser had the weight of reputation and power. Here, in a first peripheral way we see the same ever increasing pattern which will be applied again and again: incorrect science, power, reputation and money destroy good men, tarnish their reputations, corrupt good science and suppress the truth."

In 1932 Rife found the cancer virus. In 1934, he would cure cancer in humans using his frequency instrument. Rife began using Kendall's K Medium in 1931 to attempt the isolation of the cancer virus from breast tumours. The medium and microscope in this case, were not enough and a fortunate accident whereby he irradiated a sample inadvertently proved to be a decisive advantage, allowing the virus to be visualized. The incredibly small and virtually invisible structure had a breadth of one twentieth of a micron, and showed up under the microscope as purple/red in colour. The experiment was then repeated 104 times, and thus confirmed. 4 distinct forms were observed, distinct forms of the same organism. That organism, the filter passing "BX" as he called it, could reliably produce cancer in laboratory animals, as was repeated 300 times. This same organism could be transformed, depending upon the conditions and media used, into different structures found in cancer patients, a fungus, or as was later shown into bacillus coli! Pleomorphism was correct. Next he painstakingly determined the Mortal Oscillatory Rate to which the BX was attuned, and used the frequency instrument to destroy the BX. He then inoculated no less than 400 animals with filtered BX preparations, created tumours and cured those animals *over 400 times*, before attempting the first human case. Rife, was a careful and meticulous scientist.

In 1934, Dr. Milbank Johnson again gave steadfast support and was instrumental in opening the clinic in La Jolla where cancer would be first cured in the human animal. To minimize the accumulation of toxins from the dead pathogens, it was soon learned the treatment was best used for only three minutes and given at three day intervals. Of the 16 "hopeless" cases, 14 were cured in three months time. No rise in body temperature or discomfort was reported. Dr. Alvin Foord oversaw the project as chief pathologist. Chief Surgeon Whalen Morrison, George C. Dock MD, George Fisher MD, Dr. Kendall, Dr. Zite, professor of pathology at Chicago University, and Rufus B. Von Klein Schmidt President of the University of Southern California were on staff, and, Dr. Couche and Karl Meyer PhD, head of the department of bacteriological research at the Hooper Foundation, were present. Dr. Koops of the Metabolic Clinic in La Jolla signed all 14 reports and knew of the tests from his personal observation. In 1934, Royal Raymond Rife had cured cancer.

This was not an isolated fluke. Other clinics were soon opened, and more people were cured. Dr.

Couche opened a clinic and although surviving records are scant, it is clear that three patients were cured of cancer. Dr. Johnson opened a clinic of his own to reports of more success treating various conditions. Later in 1939, Dr. Richard Hamer of the Paradise Valley Sanatorium rented a frequency instrument, and proved himself adept in its use. A very organized and careful practitioner, he often treated as many as 40 patients a day, and achieved stunning results in his National City clinic, cleaning up and curing a great many conditions. Between 1934 and 1939 a very great deal of hope was brought to fruition, and unfortunately…someone noticed.

The suppression and destruction of Royal Raymond Rife—reputation, greed and our scientific dark age.

The case of an 82 year old man successfully treated by Dr. Hamer reached the powerful head of the AMA in Chicago: Morris Fishbein. The single dominant force in the AMA at that time, he was belligerent, ruthless and it is to be plainly noted that he never practiced medicine as an MD for even a single day in his life past his residency. His was a smothering and potent personality, revealing him as a political animal of the first order.

In 1939 Fishbein sent a fateful message through the California branch of the AMA seeking to buy into Rife's work and his new company, Beam Ray. Rife refused. This was the start of the end. In analyzing this supreme error, we can learn much. Rife worked with Hoyland, a self-interested and arrogant man who thought himself above his partners. He and Fishbein gathered forces. Hoyland was provided high priced lawyers. Hoyland would own Beam Ray, and Fishbein could then buy in. To refuse Fishbein, was a tragic error. When the devil knocks there is little choice but to let him in, lest he burn the house to the ground.

Rife was a patient, genteel and kind man, an intellectual, engineer and scientist who was about to enter an ugly bully's world of money, brutal assault, cruelty and blatant injustice: the American legal system. Cancer he had cured, but politics? This would soon metastasize and consume him. The opposing lawyer cut into him and attacked him. Rife was unprepared, and although the case was settled in due course in Rife's favor, the experience destroyed him on a human level...his nerves gave way. He was so deeply affected, that a doctor he trusted advised he take up strong drink, at once. This began his deterioration into alcoholism and depression. Beam Ray would be ruined, Rife shattered and the cure for cancer snuffed out. What Fishbein could not own, he would destroy. That, was that.

In a strange pair of "coincidences" the only other lab making substantial progress in the field, that of J. C. Burnett, a $250,000 lab in 1929 dollars with $500,000 invested in research and the many careful records within it, were all burned to the ground as its owner visited Rife later that year. Rife's staunch supporter Dr. Milbank Johnson would die under suspicious circumstances a few years later, perhaps poisoned according to federal inspectors, all directly before it appeared he was to announce an important breakthrough in the cure of cancer in 1944, after which his records and those of his Special Research Committee mysteriously…"vanished." Throughout this period, and afterward in the case of Rife's new partner John Crane, records and equipment were illegally seized and destroyed, and doctors intimidated into giving up the search. Dr. Couche continued the work anyway, and had his membership revoked by the AMA. Publication of one article escaped the censorship and can be seen here, as represented in the Smithsonian Annual Report, of 1944.

http://www.rife.org/magazine/smithsonian.html

Soon after its publication the report's author Dr. Raymond Seidel was shot at, a bullet slamming through his windshield as he was driving.

In the Smithsonian Annual Report of 1944 we read:

". . . disease organisms such as those of tuberculosis, cancer, sarcoma, streptococcus, typhoid, staphylococcus, leprosy, hoof and mouth disease, and others may be observed to succumb when exposed to certain lethal frequencies peculiar to each individual organism . . ."

Dr. Royal Lee of the Lee Foundation for Nutritional Research in Milwaukee spent time with Royal Rife and commented as follows:

"No medical journal was ever permitted to report on Rife's work. This one by the Franklin Institute [reprinted in the Smithsonian report] slipped by the censors, since this organization is not medical but supports general scientific activities. But that mistake was soon rectified, it appears, as there is still no general knowledge of Rife's epoch-making discoveries. Again, the iron curtain of Fishbein is effective. . . . We can give a list of various subjects on which this censorship is rigorously applied. Only the treatment of disease with synthetic drugs is carefully reported. Botanicals are played down, food as remedies are almost as taboo as Rife's work. . . the official definition of a medical remedy for disease. . . excludes automatically any vitamin, nutritional mineral or enzyme. . ."

Was there a cover up? Please take a look at this video presentation of his laboratory. Observe the massive investment and many pieces of advanced custom equipment Rife created. Observe his highly advanced work and surgical technique. Was there a cover up? Ask yourself, what happened to all of this? It appears to have vanished...without a trace.

https://www.youtube.com/watch?v=fynOk-Yldts

Cancer has been cured since 1934, and no one knows it. Fishbein, has won. When it is understood how quickly radiation machines were sold to hospitals, often with deadly result, the scope of the loss of proper testing, commercial availability and distribution of Rife's work becomes clear. Painless, three minute treatments with no side effects. Cancer, has been cured. No one knows. *Is this true?* Surely, for humanity's sake, the truth surrounding all this must come out!

The current situation and other previous substantiation.

Not surprisingly, the evidence of these facts "inexplicably" keeps showing up over and over in new work, and its imprint can be located also in history. Rife has of course, been publicly humiliated, misrepresented, ignored and discredited, although he was correct. His work is called a "myth." But the fact will not rest, for it is the fact. These…"mysteries"…abound. The cowardly suppressive reactions to them are also in evidence.

Rous had long ago discovered a cancer virus for which he would all too belatedly receive a Nobel Prize. In 1948 Dr. Virginia Livingston-Wheeler began studying tumours in which she then found the same organism. She came across the work of Dr. Eleanor Alexander- Jackson who demonstrated that tubercle bacillus went through many changes, (as Kendall, Rosenow and Rife had shown already in the 30s, a fact which had been forgotten).

Livingston-Wheeler found the extraordinary pleomorphism intriguing and wondered if cancer behaved pleomorphically. Her paper which confirmed the fact was published in 1948 in *The New York Microscopical Society Bulletin*. It concluded as follows:

"In conclusion it may be stated that a definite mycobacterium is observed in many kinds of tumours. Its presence within the tumour cells as well as within the blood of the patients suffering with the disease can be demonstrated." Livingston-Wheeler and Alexander- Jackson had demonstrated that Rous had found a virus which was *in actuality a pleomorphic bacterium.*

Livingston-Wheeler's assertions of pleomorphism were confirmed in 1950 by Dr. James Hillman of RCA labs in Princeton NJ. via electron microscopy, whose observations gave confirmation of the filtered form.

Dr. Irene Diller of the Institute for Cancer Research in Philadelphia had isolated fungus from cancerous growths in animals. Here again, was the work of Rife and Gruner, which demonstrated the changeability, the pleomorphic alterations and transformations between the BX, and a fungus. In December of 1950, *The American Journal of Medical Sciences* published Livingston-Wheeler's paper detailing how cancer cultures taken from humans and also from animals caused cancer in animal tests. Then, new cultures proved that cancer could be caused by a form of bacterium! The hurtful dismissal of this claim by Rivers and the rest, had been shown...wrong. Naturally, her discovery had all but no influence upon the situation. The cancer industry and orthodox scientific establishment had its own plans.

In 1953 Dr. Diller published her own discovery of confirmation of the transformation between fungus, and the cancer microbe: *Studies of Fungoid Forms Found in Malignancy. The Washington Post* Sep. 9, 1953 reported from Rome of their findings and ideas that: "An American research group today pictured cancer as an infectious disease. . . . 'Cancer does not consist of a localized tumour alone.' Instead, they pictured it as a generalized disease caused by an organism in the human bloodstream."

When the scientists returned home to their lab, they discovered that Dr. Rhodes of Memorial Sloan Kettering Cancer Center had shut down their work and stopped funds for the Rutgers-Presbyterian Hospital lab. The lab was shut down! Like Rife, Alexander-Jackson and Livingston-Wheeler found that their work was now...nothing. Efforts to produce a vaccine, were simply not acceptable. Surgery, radiation, chemotherapy, these ideas, *only these*, have merit and deserve funding. Inexpensive cure was not on the cards. A new approach was not acceptable. Money. Reputation. This is modern medical practice and research. This...is tragedy.

In 1983, Livingston-Wheeler wrote in her book, *The Conquest of Cancer:*

"Because of the suppressive actions of the American Cancer Society, the American Medical Association and the Food and Drug Administration, our people have not had the advantage of the European research.

This work has been ignored because certain powerful individuals backed by large monetary grants can become the dictators of research and suppress all work that does not promote their interest or that may present a threat to their prestige."

She demonstrated that cancer is not as accepted orthodox science proposes. In the New York Academy of Sciences report from October 30 1970 we read:

"Microorganisms of various sorts have been observed and isolated from animal and human tumours, involving viruses, bacteria, and fungi. There is, however, one specific type of highly pleomorphic microorganism that has been observed and isolated consistently by us from human and animal malignancies of every obtainable variety for the past twenty years. . . that organism has remained an unclassified mystery, due in part to, its remarkable pleomorphism and its stimulation of other microorganisms. Its various phases may resemble viruses, micrococci, diptheroids, bacilli, and fungi."

Today a music professor turned cancer researcher named Anthony Holland is shattering cancer cells with targeted frequencies. He was inspired by the work of Rife. His approach is aimed at the cancer cells themselves. Will he succeed? Will the cells to be shattered cause toxic shock in the body? Does the source pathogen, the pleomorphic BX need to be targeted? Let us look to this fine man, and see if he can find the answers for us.

Now Luc Montagnier, the scientist who had won a Nobel Prize for his discovery of the AIDS virus, has found some strange effects whereby filtered and presumably sterile preparations, appear to reconstitute disease related organisms in cultures, and hypothesized with good supporting evidence the existence of aqueous nanostructures of a mysterious sort, offering us also in other work a glimpse into the quantum aspects of what possibly appears to be the mechanics of pleomorphic transformations. Naturally, he is accused of poor lab technique and contamination in his experiments. Do you believe this Nobel Prize winner cannot prepare a proper experiment? What is happening here, why not look into the mysterious aqueous nanostructures, so akin to the BX or perhaps some quantum informational part of pleomorphic processes? Instead, his reputation is tarnished and insults abound. Can you see it? If he is right, many diseases, not just cancer but perhaps Alzheimer's, Parkinson's, Multiple Sclerosis, Rheumatoid Arthritis and a great many more may be treatable, or even curable. Does this seem familiar? Can you see it? We must discover if this is true.

It should be remembered also that dismissal of the pioneering work of Rife is not confined solely to America. On the web page of Cancer Research UK, reference is found to Rife and his work but most of the statements contained therein are erroneous. Firstly, and possibly most importantly, Rife's work and all his results were based on the use of his own microscope and his own Frequency Instrument – not on the so-called Rife Ray machines on sale today. Again importantly it should be noted that all the talk on this site refers to the Rife machine destroying cancer cells but, in truth, the original claim of Rife was that his machine enabled him to destroy the microbe which caused the production of cancer cells and, having destroyed the microbe, the cells eventually returned to their original healthy state. They also claim that the Rife approach has not been through the usual process of scientific testing and, in a sense this is true but, as shown above, such a statement is nowhere near the real truth. Rife's approach was meticulously tested and was open to further, more public, testing but this was maliciously denied and his work is now largely forgotten.

In 2007, in the book *Exploding a Myth*, the whole question of the existence and influence of these same destructive influences operating in the discipline of physics was explored through several examples. The restrictions placed on true open minded original thought were noted and it was pointed out that, while such problems in pure physics might not concern many people, if such restrictions did occur in physics then they would occur in other branches of science and, if the branch involved was medicine, then it was putting peoples' lives at stake. It is tragic to reflect that the facts outlined above indicate the correctness of that earlier prophesy.

Recent ideas seem to indicate that it might be possible for drugs, so toxic and profitable, to be replaced with safe quantum information. Is there any real value in this idea? Can the multitude of horrible diseases be cured by spotting the common pleomorphic mechanics of their replication, and addressing the problem with inexpensive painless treatments? Can we redo today what Rife apparently accomplished so long ago? These are the questions of the future.

Our research has unearthed a distinct and disturbing possibility:

"Cancer has been cured since 1934."

For the sake of humanity we ask aloud:

Is this true?

This article owes a clear and substantial debt to Barry Lynes' excellent book: *The Cancer Cure That Worked.*

5. Thoughts on Redshift and Modern Cosmology.

Introduction.

The 2008 book *Facts and Speculations in Cosmology* by Jayant Narlikar and the late Geoffrey Burbidge [1] ends with the query

Do we really understand the nature of the redshift?

This query is all the more devastating when one considers the central position the notion of redshift occupies in astrophysics, astronomy and cosmology. These three fields of scientific endeavour are cited separately here to emphasise the fundamental importance of the stated query. The whole notion of redshift is central to so many aspects of these three areas of scientific endeavour and it is worth reflecting further on this comment for a moment. It is redshift which is an important factor behind so much of our determination of distance in the Universe; it is redshift which is behind the idea that our Universe is expanding; it is redshift which is a factor in the introduction of the notion of so-called dark matter. A moment's contemplation indicates that this list, though short, is indeed formidable and even the three examples cited here bring instant realisation of the importance of the concept of redshift in modern science but, for the present purpose, attention will be restricted to the first example, the determination of distance in the Universe.

Some comments on redshift.

As is well known, there are several possible contributions to the redshift observed on any one occasion but the one on which attention is usually focussed is that due to the so-called Doppler effect – a frequency change in waves occasioned by relative movement between source and observer, with a decrease in frequency, or redshift, indicating movement of the source away from the observer. Hence, the redshift is inextricably linked to the motion of the source but, following the work of Hubble in the first quarter of the twentieth century, a further link became apparent. Hubble had been working for some time estimating the distances to various galaxies when he realised that the higher the value of the radial velocity of a galaxy as indicated by its redshift, the farther away it was according to the distance determination methods used. This observation led eventually to the establishment of the well-known distance – redshift relationship, which has proved so useful over the years. Originally the relationship applied to nearby regions but has been assumed to hold for all of our Universe. Considering the problems facing investigators in examining the Universe, this is not really an unreasonable assumption to make but it is still an assumption and so, could be false. Of course, much the same is true of assuming Newton's laws applicable throughout the Universe and that assumed validity has, in fact, been challenged

However, whether or not it is valid to extend regions of validity in this cavalier manner, the biggest challenge to the relationship possibly arises through major questions concerning the interpretation of the observed redshift.

The observations of Halton Arp.

One important class of objects to be considered in the present context is provided by the quasars; the most 'distant' quasars are thought to have redshifts far in excess of those for the furthest galaxies. It is accepted by many that there were far more quasars and, indeed, radio galaxies in the past than there are now. However, this whole question is, or should be, a completely open one. Many seem to give the impression that everything in this area is absolutely clear cut and anyone opposing the generally accepted view is to be ignored as lacking in understanding of the truth. Frankly this appears to be the view adopted in the corridors of conventional wisdom towards the work and ideas of Halton Arp. While able to make use of the most powerful of telescopes, Arp also discovered that many pairs of quasars which possess extremely high redshift values appear to be associated physically with galaxies having much lower redshift values; galaxies, in fact, which are known to be much closer to the earth than the redshift values of the quasars concerned would imply. This all follows from the Hubble law which indicates that objects having high redshift values must be receding from the earth very quickly and, therefore, must be found at large distances from the earth. Hence, Arp was faced with the intriguing question of how objects with totally different redshift values, objects which according to 'conventional wisdom' had to be located at totally different distances from the earth, could be physically associated – in some instances, Arp's photographs actually showed a physical bridge between the quasars and the associated galaxy. As has been recorded many times, Arp has many photographs of pairs of quasars, with high redshifts, symmetrically located on either side of low redshift galaxies. It has to be noted that these pairings occur far more often than the probability of random placement would allow. Of course, the main problem with Arp's photographs is that according to orthodox theorists, high redshift objects must be at a great distance from the earth; to them high redshift is effectively a measurement of distance from the earth. It is often claimed by the advocates of 'conventional wisdom' that Arp's statistical analysis is in error; after many years, this still seems to be the main line of attack on his work. However, from all the accumulated evidence it seems there is no satisfactory foundation for criticising Arp's work on the basis of the statistics involved, and that seems to be the only criticism actually offered. Much of Arp's work is well-documented in his book *Seeing Red* [2] and reference should be made to this work for further details of the specific points involved.

As might be expected, this work of Arp's has not been welcomed by the orthodox astronomical community because, if accepted, it casts severe doubt on the assumption, which is quite basic to most, if not all, of accepted cosmological theory, that objects possessing a high redshift must be far away from the earth. Whether people approve of his work or not, it is undoubtedly true that Arp's work raises serious questions about the present state of cosmological theory and to ignore these questions, as some would advocate, should not be an option for any serious investigator in the field. It is also undeniably true that serious questions about the true interpretation of observed redshifts remain and must be addressed with open minds if real progress is to be made. However, a new method to determine distances in space has been announced recently [3]. This proposal, involves the possible use of quasars as standard candles. Once again, though, the concept of redshift appears central to the discussion. Hence, interesting and valuable as this reported work may be, it does seem its real usefulness will depend on a correct understanding of redshifts observed. Therefore, all in all, it is seen that probably the most important question facing cosmology concerns the correct meaning of this concept of redshift – something which initially seemed so simple to interpret.

On top of this, though, various other questions have arisen but the importance of these has frequently been played down by the scientific press. Most of these depend on observations of magnetic fields and electric currents in space and one problem with many is that, in all likelihood, the experiments proposed to help solve the associated problems have been carried out and documented already.

Magnetic fields in space.

On 5th December, NASA announced that its Voyager 1 spacecraft had entered a new region between our solar system and interstellar space [4]. In this announcement, one of the more interesting comments is that "Voyager has detected a 100-fold increase in the intensity of high-energy electrons from elsewhere in the galaxy diffusing into our solar system from outside". This comment is of interest because, apart from the word 'diffusing', it describes what the electrical model of our universe expects in the virtual cathode region of the solar discharge boundary.

Also, on 6th December, it was revealed that a new all-sky map shows the magnetic fields of the Milky Way with the highest precision [5]. It was claimed that the origin of galactic magnetic fields remains unknown despite intensive research, although it was seemingly assumed that they are constructed via dynamo processes such as are said to occur – in violation of Cowling's well-known theorem incidentally – in the interiors of the Earth and the Sun.

Some years ago, in an entirely different context, Sir Winston Churchill advised people to learn from the lessons of history and, in the present context, it might seem appropriate to follow this advice in astrophysics. Hence, in this spirit, it might be noted that, following the introduction of Newton's mechanical ideas, work still proceeded apace investigating electromagnetic phenomena and this continued at least into the earlier years of the twentieth century, as is evidenced by the contents of J. J. Thomson's book *Electricity and Matter* [6]. However, this book provides but one example to illustrate the very real emphasis on work involving the effects of the electric and magnetic fields, work which, incidentally, constantly sought an explanation for the concept of mass in terms of those forces. However, after those early years of the century, the emphasis seems to have shifted to explanations of phenomena purely in terms of gravitational effects. Considering that it is accepted that much of the matter in the universe is in the form of plasma, this might be thought a retrograde step. One may only speculate as to why the emphasis of much scientific research changed in this way. However, thanks to people like Birkeland, Alfvén and, more recently, Peratt, work in the areas of electromagnetism and plasma physics has continued.

The work on plasmas and other electromagnetic phenomena has inspired people to examine astronomical phenomena in these terms and this has resulted in the so-called Electric Universe idea as expounded, for example, in the books *The Electric Universe* [7] and *The Electric Sky* [8]. Reading through this material makes one immediately aware that just like accepted theory the electric universe ideas are supported by computer modelling, but it is also able to draw on parallels between astronomical phenomena and plasma phenomena observed in the laboratory. Admittedly, drawing such parallels involves scaling up tremendously but assuming this possible is little different from assuming that laws seemingly applicable here on the Earth are also applicable in the Solar System and, indeed, throughout the universe. At least visually, some of the phenomena observed in the laboratory are very like what is observed by some of the most powerful of telescopes. Electric currents in plasma naturally form filaments due to the so-called 'pinch effect' of the induced magnetic field. Electromagnetic interactions cause these filaments to rotate about one another to form a helical 'Birkeland Current' filament pair and this is very much the structure seen in the Double Helix nebula near the galactic centre; again, the Hubble image of the planetary nebula NGC6751 looks remarkably like the view down the barrel of a plasma focus device. Examples such as these prove nothing but might awaken people to the possibility of alternative explanations for at least some astronomical phenomena.

The Heritage of Kristian Birkeland.

Much of the laboratory work originated with the work of Kristian Birkeland more than one hundred years ago. It was during his Arctic expeditions at the end of the 19th century that the first magnetic field measurements were made of the Earth's polar regions. His findings also indicated the likelihood that the auroras were produced by charged particles originating in the Sun and guided by the Earth's magnetic field. Birkeland, though, was an experimentalist and is still known for his Terrella experiments carried out in a near vacuum and in which he used a magnetised metallic sphere to represent the Sun or a planet and subjected it to electrical discharges. By this means, he was able to produce scaled down auroral-type displays as well as analogues of other astronomical phenomena. These claims, however, were only vindicated finally by satellite measurements in the 1960's and 70's. To that point in time, his experimental and observational achievements had tended to be overshadowed by the purely theoretical predictions and explanations of the geophysicist, Sydney Chapman. Powerful mathematics seems to have held sway over the more expected techniques of physics – experimentation and observation, with mathematics a mere tool to be used when necessary. This is not to decry Chapman's work but to emphasise the overwhelming importance of the physics when investigating natural phenomena.

Birkeland also showed experimentally that electric currents tend to flow along filaments shaped by current induced magnetic fields. Of course, this confirmed observations of Ampère that indicated that two parallel currents flowing in wires experience a long range attractive magnetic force that brings them closer together. However, as plasma currents come closer together, they are free to rotate about each other. Such action generates a short range repulsive magnetic force which keeps the filaments separated so that they are, in effect, insulated from each other and able to maintain their separate identities. The end effect is for them to appear like a twisted rope and it is this configuration which is termed a 'Birkeland current'. Satellites orbiting above the auroras in the 60's and 70's were able to detect a movement of ions, indicating that electric currents were present. Later missions found quasi-steady electric fields above the auroras following the magnetic field lines, thus lending some credence to Birkeland's claim of the existence of an electric circuit between the earth and the Sun.

However, the so-called Electric Universe is really just a hypothesis, a new way of interpreting known data by using both new and well-established knowledge relating to electricity and plasma. It should be emphasised immediately that, in this new interpretation, gravity still has a role to play but it is a secondary one since the electric force is so much more powerful. A major point to be stressed from the outset is that, in this interpretation of astronomical phenomena, scientists are able to call on evidence from laboratory based experiments to help form and support suggested explanations for a wide variety of phenomena. It has been found that, as explained in more detail in the above-mentioned books, a plasma in a laboratory is a good model for providing possible explanations for many recently observed astronomical phenomena which, in several cases, have puzzled astronomers seeking explanations via more usual routes. This is not to say that gravity is ignored and regarded as irrelevant; rather, the possible effects of the electromagnetic force on astronomical phenomena are investigated while still recognising the importance of gravitational effects. In the electric universe, the gravitational systems of galaxies, stars, moons and planets are felt to have their origins in the proven ability of electricity to generate both structure and rotation in plasma. It is felt further that the force of gravity assumes importance only as the electromagnetic forces approach equilibrium. As has been noted already, great consternation has been caused in astronomical circles by the realisation that gravity, as presently understood, cannot explain much that is observed if the amount of mass available is as now felt to be present. Hence, instead of positing the existence of 'dark matter' or following the path of modifying Newton's well-tried law of gravitation significantly, it is suggested here that the effects of the electromagnetic force be examined to see if, in conjunction with orthodox ideas on gravity, these puzzling observations can be explained. However, returning to the realisation that much of the matter permeating the Universe is in the form of plasma, it might be remembered that these clouds of plasma respond to the well-known laws of Maxwell. Also, as pointed out by Scott in his book [8], another law, formulated by Lorentz, does help explain the galactic speeds alluded to earlier. This law states that

a moving charged particle's momentum (speed or direction) can be changed by application of either an electric field or a magnetic field or both.

This seems a highly likely contributory factor, at least, causing galaxies to rotate as they are perceived to do but would indicate, contrary to the accepted view, that gravity has less to do with things than has been thought. However, it should be noted that nowhere is it being suggested that Newton's law of gravitation is in error; it is simply being suggested that, in deep space where everything swims in a sea of plasma, the Maxwell – Lorentz electromagnetic forces dominate over those of gravity.

It might be remembered also that the Lorentz force alluded to here changes a charged particle's momentum and that change is directly proportional to the strength of the magnetic field through which the particle is moving. Further, the strength of a magnetic field produced by an electric current is inversely proportional to the distance from the current but the gravitational force between stars is inversely proportional to the *square* of the distance. This well-known difference between the two forces could lie at the heart of the problem of the galactic rotation curves; certainly it seems an avenue worth exploring further, especially considering the fact that more and more space missions are indicating that electromagnetic forces are distributed more widely throughout space and are, of course, many orders of magnitude stronger than gravitational forces.

As well as a great many laboratory experiments being performed to establish plasma properties [9] , it has been shown also, using the Maxwell and Lorentz equations, that streams of charged particles, such as are found in the intergalactic plasma, will evolve into the familiar galactic shapes under the influence of electromagnetic forces. The results fit extremely well with the observed velocity profiles in the galaxies and all this without recourse to missing mass or other esoteric entities. Much of this simulation work has been carried out by Anthony Peratt and is reported in various issues of the IEEE Transactions on Plasma Science. However, recent reports [10] of the production of magnetic fields in a laboratory by using a high-power laser to explode a carbon rod in helium gas in an effort to simulate the plasma out of which the first galaxies are thought to have been formed are accompanied by discussion of experiments to be performed to help examine these phenomena in laboratories here on earth. It seems that the hope is to examine the physics of the cosmos over billions of years in a laboratory here on Earth. Such experiments as proposed by more than one group in the United Kingdom usually involve examining plasmas. The grave suspicion must arise that such experiments have often, if not always, been carried out already by the likes of Peratt and his mentors. It is worrying that so many either do not know of this body of work or dismiss it partially because the journal in which many of the results are published is regarded by some as non-prestigious, although how such a comment can be made – apparently seriously – about the above-mentioned IEEE journal remains something of a mystery.

References.

[1] J. Narliker & G. Burbidge, 2008, *Facts and Speculations in Cosmology*, C.U.P., Cambridge.

[2] H. Arp, 1998, *Seeing Red,* Apeiron, Montreal.

[3] '*Quasars shine a new light on cosmic distances*', Science Daily, May, 4th, 2012.

[4] http://www.jpl.nasa.gov/news/news.cfm?release=2011-372

[5] http://www. physorg.com/news/2011-12-all-sky-magnetic-fields-milky-highest.html

[6] J. J. Thomson, *Electricity and Matter*, Westminster: Archibald Constable & Co., 1904.

[7] W. Thornhill and D. Talbott, *The Electric Universe*, Portland, Mikamar Publishing, 2002.

[8] D. E. Scott, *The Electric* Sky, Portland, Mikamar Publishing, 2006.

[9] A. Peratt, *Physics of the Plasma* Universe, New York, Springer-Verlag, 1992.

[10] '*Laser hints at how Universe got its magnetism*', Science Daily, March 24th, 2012.

6. Thoughts on the 'Big Bang'.

Introduction.

On the very first page of his book *Before the Beginning*, the Astronomer Royal, Lord Rees, states categorically that "Our universe sprouted from an initial event, the 'big bang' or fireball'". What a truly amazing statement with which to begin any book or piece of writing. However, is it true?

The big bang as a valid model of the Universe has been under close scrutiny almost since it was proposed and many of the queries concerning it remain. These queries tend to be 'swept under the carpet' but in a rather subtle way. The rise of popular science books has provided a means whereby the general public is persuaded to believe in the ideas accepted as founding 'conventional wisdom'. This has been supported by a proliferation of carefully constructed, well presented public lectures. The 'solutions' to various problems are presented as indisputable facts; the notion that other possible explanations exist is carefully suppressed or, in the case of the steady state theory, mildly ridiculed.

The whole idea of the big bang goes back to the theoretical investigations of Alexander Friedmann and Georges Lemaître [1] in the earlier years of the last century following Einstein's publication of his General Theory of Relativity. Its movement to a position of prominence, if not pre-eminence, in cosmology might be felt to have been brought about by its eloquent advocacy at the hands of George Gamow [2] in the mid to late 1940's, ably supported by such as J. Robert Oppenheimer. It is quite widely claimed that the standard big bang model makes three major predictions which have been verified observationally. If that were true beyond all reasonable doubt, it would indeed be a theory to take very seriously. However, are these claims unquestionably true? First, it is claimed that the model predicts distant galaxies receding from one another at speeds proportional to the distance between them. This view is supposedly supported overwhelmingly by Hubble's discovery of the redshift of light from celestial objects in the 1920's. Secondly, the model is claimed to predict the existence of background radiation which is seen as a remnant of the original big bang. Support for this comes from the detection of the cosmic background radiation by Arno Penzias and Robert Wilson in 1965 [3]. Some also claim that the relatively recent examination of the properties of this background radiation by the COBE satellite again confirm totally the predictions of the big bang. Thirdly, the model is said to predict successfully the abundances of the light elements such as helium, deuterium and lithium. At the same time, these claims are taken to imply that no other theory can explain these phenomena and there are no doubts about these deductions from the basic idea of the big bang. It goes almost without saying that the interpretation of experimental and observational results which leads to confirmation of the 'truth' of the big bang theory is accepted without question. However, is the situation quite as clear cut as that? Are all the questions answered, and answered both successfully and correctly?

As far as modern ideas are concerned, one of the first major advances came with Hubble's evidence that three nebulae, M31, M33 and NGC6822, were to be found at distances far beyond the remotest parts of our own galaxy. It was accepted that these were totally separate from the Milky Way. Not long after establishing that these nebulae were extragalactic systems, he also showed that the redshift of their spectral lines increased with distance. Utilising the most obvious interpretation of redshift, that is that it is a Doppler shift occasioned by the recession of the source, it is easily seen that Hubble's result may be taken to indicate that the Universe is expanding and the most distant galaxies are receding fastest. By looking at things in reverse, this is seen to mean that the Universe was much denser in the past and there is a tendency to extrapolate back to claim that, at some distant time, all the matter in the Universe was so highly compressed that it was all confined to a single point! At this point the *'Cambridge Encyclopædia of Astronomy'* comes into its own as far as fair, scientific examination of this issue is concerned. It claims that great care should be taken, since, "it is possible that the simple interpretation of the redshift is not correct, and that the expansion is illusory." Even if the fact of expansion is accepted, "it does not necessarily follow that the Universe was denser in the past than now, for implicit in that conclusion is the assumption that matter in the Universe is neither created nor destroyed." However, it is pointed out also that the hypothesis that the redshift is a Doppler shift occasioned by recession of the galaxies is acceptable scientifically since it is consistent with the known laws of physics. Reflecting the time of writing, it is claimed that "no other scientifically acceptable hypothesis has yet been proposed" but it does note that, as far as the position existing at that time was concerned, there was no proof that that was the true explanation. The encyclopædia article continues by noting that, since the time of Hubble's original hypothesis, many more observations had been made which served to confirm his postulated relationship between distance and velocity of recession. It is claimed that no obvious deviations from the simple linear relationship,

Velocity =

Hubble parameter x distance in megaparsecs,

have been detected.

Hubble also spent a considerable amount of time investigating the distribution of galaxies in the Universe. Obviously, such observations were restricted by the instrumentation available but, nevertheless, he noted that, on very large scales, the Universe does appear homogeneous; there is no obvious sign of diminution of numbers of galaxies as the accessible limits of the Universe are approached. Also, the Universe was found to look more or less the same in all directions and the cosmic expansion seemed to be proceeding at the same rate in all directions; that is, the Universe is said to be isotropic. All this is taken to mean that there is no meaningful centre for our Universe and as confirmation that our own galaxy, the Milky Way, certainly occupies no privileged position within the Universe. Strong confirmation for the isotropic nature of the Universe is felt to be provided by the so-called cosmic background radiation, a component of radiation found by radio astronomers which is itself isotropic to a very high degree and is inexplicable as noise within receiving systems or as originating from any known radio sources. This radiation is, of course, that background radiation mentioned earlier. Since Hubble's time, however, observing equipment has changed for the better and systems are now observed quite regularly which emit far more radiation than many of those observed by Hubble. One important class of objects to be considered here is provided by the quasars; the most 'distant' quasars are thought to have redshifts far in excess of those for the furthest galaxies. It is accepted by many that there were far more quasars and, indeed, radio galaxies in the past than there are now. This, if true, implies that, in the past, the Universe was different from now and this seems to pose a serious problem for the Steady State Theory of the Universe, as well as offering extremely strong support for alternatives, especially the big bang model. However, this whole question is, or should be, a completely open one. Many seem to give the impression that everything in this area is absolutely clear cut and anyone opposing the generally accepted view is to be ignored as lacking in understanding of the truth. Frankly this appears to be the view adopted in the corridors of conventional wisdom towards the work and ideas of Halton Arp.

The Work of Halton Arp.

While able to make use of the most powerful of telescopes, Arp also discovered that many pairs of quasars, or more correctly quasi-stellar objects, which possess extremely high redshift values appear to be associated physically with galaxies having much lower redshift values; galaxies, in fact, which are known to be much closer to the earth than the redshift values of the quasars concerned would imply. This all follows from the Hubble law which indicates that objects having high redshift values must be receding from the earth very quickly and, therefore, must be found at large distances from the earth. Hence, Arp was faced with the intriguing question of how objects with totally different redshift values, objects which according to 'conventional wisdom' had to be located at totally different distances from the earth, could be physically associated – in some instances, Arp's photographs showed a physical bridge between the quasars and the associated galaxy. As has been recorded many times, Arp has many photographs of pairs of quasars, with high redshifts, symmetrically located on either side of low redshift galaxies. It has to be noted that these pairings occur far more often than the probability of random placement would allow. Of course, the main problem with Arp's photographs is that according to big bang theorists, high redshift objects must be at a great distance from the earth; to them high redshift is effectively a measurement of distance from the earth. It is often claimed by the advocates of 'conventional wisdom' that Arp's statistical analysis is in error; after many years, this still seems to be the main line of attack on his work. However, from all the accumulated evidence it seems there is no satisfactory foundation for criticising Arp's work on the basis of the statistics involved, and that seems to be the only criticism actually offered. Much of Arp's work is well-documented in his book *Seeing Red* and reference should be made to this work for further details of the points involved.

As indicated, this work of Arp's has not been welcomed by the orthodox astronomical community because, if accepted, it casts severe doubt on the assumption, which is quite basic to big bang theory and, therefore, to most if not all of accepted cosmological theory, that objects possessing a high redshift must be far away from the earth. Hence, all the claims of the big bang model which depend on the orthodox interpretation of redshifts must be re-examined. Again, Arp's hypothesis, backed by such eminent physicists as Hoyle, Burbidge and Narlikar, casts doubt also on the notion that black holes lurk at the centre of quasars. However, no black hole has yet been identified beyond reasonable doubt, but, if one did exist, it is assumed that it would be drawing matter to itself rather than ejecting it at very high velocities. So once again, Arp displeases the establishment by proposing a solution to a very real problem which suggests matter being ejected from a central mass rather than absorbed into it. In much current astronomical literature, there seems to be a preoccupation with the death of stars and, in some ways more importantly, with the colliding or merging of galaxies. Arp's view, and one supported by Hoyle and many of his associates, is that it is rather the birth of galaxies that is being witnessed; instead of viewing and contemplating possible collisions, it is rather separations that are being seen. It might be felt that this view is more in keeping with big bang cosmology in that the big bang supporters claim the universe to be expanding and so, everything should be moving farther and farther apart; collisions, it would seem, should be highly improbable occurrences. However, this view is too simplistic and absorbing actions, such as that envisaged by black holes, are readily incorporated into big bang theory. The Arpian view of what is happening is taken to be in direct opposition to the big bang theory, probably because it may be interpreted as implying creation of matter and this notion is, of course, at the heart of the new quasi-steady state theory of Hoyle and his collaborators [4], as well as being seemingly contrary to well-established conservation laws. This quasi-steady state theory is a modification of the old steady state theory proposed by Bondi, Gold and Hoyle [5] many years ago and is a modification proposed in answer to criticisms of the original. It might be argued that they have listened to their critics and attempted to provide an answer. The difference between this modification and changes made to the big bang theory is that, in this case, it seems that the theory was modified but, in the case of the big bang, it seems that, when a problem is pointed out, something is simply added on in an attempt to solve that immediate problem.

As indicated above, at one point in time - actually by about 1950 - there were really two rival theories attempting to explain the origin and workings of the Universe. These were the big bang model and the steady state theory. Both accepted the idea that the Universe was homogeneous, isotropic and was expanding against the pull of gravity. However, the steady state theory assumed that matter could be both created and destroyed spontaneously, whereas the big bang did not. The idea of spontaneously creating or destroying matter challenges widely, and strongly, held views on conservation and so will be anathema to many. On the other hand, one apparently awkward consequence of the big bang is that, at some time in the distant past, all matter seems to have been concentrated in some state of infinite density; that is, a singularity, the cosmic singularity, existed. It is often claimed that this singularity is a serious defect in the big bang theory on philosophical grounds but, in many areas of mathematics and physics, it is more usual to note that a singularity heralds the breakdown of a theory or that there are limits to the range of applicability of a particular theory. It is interesting to realise that, for some reason, no such restriction is imposed in this case or, indeed, in the case of black holes of the type which are said to emerge via the general theory of relativity. In both these cases, attempts are actually made to give physical meaning to mathematical singularities. Apparently, it is this singularity in the case of the big bang which prompted Bondi, Gold and Hoyle to propose the steady state theory in which matter could be created spontaneously at a rate which compensated the reduction in density brought about by the cosmic expansion. Such a Universe would presumably have no beginning or end, it would have both an infinite past and future, but, possibly more importantly, the model would have no singularity.

Considering this latter point concerning the steady state theory, it is interesting to wonder at the possible role played by fundamentalist religion in the seemingly widespread acceptance of the big bang model and the resultant rejection of steady state theory. A moment's reflection indicates that the possibility of such a link is not totally ludicrous. If one considers the first nineteen verses of the King James version of the Bible, the first obstacle to be overcome is the unscientific language used. However, when that is done, it becomes immediately apparent that one valid interpretation of what appears in print is that the Universe was created quite suddenly, spontaneously in fact. The ordering that follows also links quite well with big bang philosophy. It might be argued, quite reasonably, that light would be necessary before grass and fruit trees could exist but, bearing in mind that the ideas, or stories, of Genesis are extremely old and may be interpreted sensibly only as representations produced by people without the benefit of modern scientific knowledge to illustrate, to a scientifically uneducated people, the beginnings of the Universe and of life on earth, the correspondence with the ordering of events according to the big bang theory is remarkably close. It might be noted specifically that even the presence of radiation before the formation of the stars may be inferred from verses fourteen to nineteen inclusive. However, was Genesis ever intended to be taken literally? Was it ever meant to be the literal truth describing the origin of the Universe and life in it? On this question, as with questions of theories of evolution, various views abound. Amongst these, is the view that the answers to the above two questions are in the affirmative. There are, and always have been, people who do believe the book of Genesis to be literally true. Some of these people are, and have been, serious scientists. This may seem almost a contradiction in terms but it is, nevertheless, true. It is, therefore, not difficult to see precisely how the big bang theory will appeal to such people as being the perceived 'Word of God'. It is very easy, but also very unfair, to ridicule such a standpoint, since the obvious temptation is so strong.

This short semi-religious discussion merely serves to raise another question and that is whether, or not, religious fundamentalism played any part – however small – in the acceptance of the big bang theory over the steady state theory? Indeed, it is not unreasonable to wonder if, with the seeming resurgence of religious fundamentalism in present day society, it is one factor keeping the big bang theory so much to the fore. However, whatever the reason, it is still the case that the validity of the big bang theory seems accepted totally without question by much of the world-wide scientific community.

Cosmic-Microwave-Background Temperature.

One of the most vociferous of early proponents of the big bang theory was George Gamow. He and Ralph Alpher first put the theory forward seriously in 1948 and almost immediately became engaged in a war of words with the supporters of the Steady State Theory. However, Gamow's theory did, apparently, make some important predictions. Namely that there should be an abundance of helium of about twenty-five per cent by mass, and that it should be possible to observe the remnants of the radiation from the early hot phase of the universe's existence and this should be an isotropic radiation field with a black body spectrum with a temperature of a few degrees. The estimates forwarded for this temperature varied, however, between about five degrees and fifty degrees absolute. This was interesting because, as early as 1926, Sir Arthur Eddington [6] had predicted a temperature of space of three degrees absolute, purely on the basis of the radiation received from the stars. This calculation is very crude but the magnitude of the result provides food for thought, if nothing else. Again it might be noted that Eddington was discussing the temperature of interstellar space due to stars in our own galaxy; he was not considering intergalactic space. However, be that as it may, the big bang theory received possibly its biggest boost, both within and without the scientific community, with the discovery by Penzias and Wilson of the cosmic background radiation, - that background radiation which is almost universally recognised nowadays as a left-over of the original big bang. Here it is important first to ask whether or not this discovery of the cosmic background radiation is, in fact, really due to Penzias and Wilson. It must be acknowledged that the existence of this background radiation was not universally recognised at the time of Penzias and Wilson. However, its existence had been detected in the late thirties and early forties by various astronomers. In 1941, McKellar had interpreted the observed data and had shown it to be caused by radiation excitation, which was taken to be black body and the temperature required for the observations to be properly explained was found to be 2.3°K. Hence, the detection of the microwave background should more correctly be dated from 1941. It is, in all fairness, understandable that this did not happen. In 1941, the world was in turmoil at the height of the Second World War and McKellar's important work did not appear in a front line journal. However, the truth has been known for some time now. Hoyle, in particular, has not been backward in publicising its existence. It is to be hoped that McKellar will soon be given the credit he surely deserves.

It might usefully be noted that, long before Gamow and others began to espouse the Big Bang theory, several notable scientists had followed the lead of Guillaume and Eddington and proposed estimates of the temperature of intergalactic space. Following initial work by Millikan and Cameron in which it was deduced that the total energy of cosmic rays at the top of the atmosphere was a tenth of that due to the heat and light emitted by the fixed stars, Regener eventually concluded that both energy fluxes should possess more or less the same value. In an article of 1933, he used this as a basis for deducing a value of $2.8^{\circ}K$ as the temperature characteristic of intergalactic space. This work was discussed favourably by no less a person than Walther Nernst who, by 1912 had developed the notion of a stationary state universe. By 1937, he had further developed this and actually proposed a 'tired light' explanation for the cosmological redshift; that is, he suggested that absorption of radiation by an aether which decreased the energy and frequency of galactic light. Whether one accepts or rejects these ideas now, it should be noted that, in all these separate pieces of work, as well as in subsequent examinations by such as Max Born, Stefan-Boltzmann's law, which is characteristic of black body radiation, is of paramount importance. Also, in none of this work, nor in that of McKellar, is any reference made to the big bang theory; it is simply not necessary to introduce it in order to achieve the results cited!

However, nowadays it is the papers by Gamow and by Alpher and Herman, dating from 1948 [7], which tend to hold pride of place where discussion of the background temperature is concerned. They pointed out that, if helium was synthesised in the early universe, then, in present times, there should exist a radiation field with a temperature of approximately $5^{o}K$. Gamow offered another prediction of the temperature of the background radiation in his 1952 book *The Creation of the Universe*, but this time the estimate, which was claimed to be "in reasonable agreement with the actual temperature of interstellar space", was roughly $50^{o}K$. Nevertheless, it is frequently claimed that Gamow and his collaborators predicted the $2.7^{o}K$ temperature (even though their lowest estimate was in fact $5^{0}K$) before the 'discovery' of Penzias and Wilson, whereas the steady state theory did not. This was, and is still, hailed as one of the strongest arguments in favour of the big bang model. However, it must not be forgotten that the original steady state theory did not rule out the existence of a background radiation and, as is pointed out in Hoyle's last book, some unpublished calculations by Hoyle, Bondi and Gold, dating from about 1955, indicated a temperature associated with that radiation of $2.78^{o}K$. Obviously, revealing this at the time of the supposed discovery of the cosmic background radiation would have produced totally the wrong reaction. However, it is important to note that, from the outset, the steady state theory never ruled out the possibility of there being a background radiation in existence. Therefore, it is obviously totally incorrect to use the existence of this background radiation as a major reason for attempting to discount that theory.

The Synthesis of Helium.

When introducing the articles by Gamow and by Alpher and Herman above, it was noted that they made reference to the synthesis of helium in the early universe. They were using this to support their claim that, if this were so, a radiation field pervading the whole of space should exist now. This, of course, raises the entire question of the process behind the synthesis of helium and the other light elements. It is of interest to realise that, once again, the papers referred to here were not the earliest attempts to raise this problem. Actually, the earliest article by Gamow appeared in 1946 [8]. In it, he argued that, in the early universe, the chemical elements were synthesised by neutron addition. Hoyle also produced his first article on stellar nucleosynthesis in 1946 [9] and, interestingly, his view was the direct opposite of that proffered by Gamow. In fact, it is quite widely accepted now that the originator of the theory behind the synthesis of the light elements was Hoyle and a great many people are still puzzled by the fact that he received no part of the Nobel Prize awarded for that work. While the overall thesis of this present work is concerned with the place of accepted 'conventional wisdom' in the scientific world, this treatment of Hoyle inevitably raises the spectre of 'politics' within the scientific establishment. However, let us now return to the articles by Gamow and by

Alpher and Herman. After one or two early hiccups, Gamow and his collaborators produced a theory whose key point was the essential requirement that an amount of helium be synthesised in agreement with the observed value of approximately 0.25 by mass when compared with hydrogen. It might be noted immediately that this fraction is not thought to be constant in time and that alone raises questions. Although it is known that helium is produced from hydrogen in the interior of stars, it was always felt, and still is, that stellar synthesis would make only a negligible contribution to this observed fraction. As has been pointed out by Hoyle and his collaborators, it would take of the order of 10^{11} years to increase the value of this fraction from zero to 0.25. Around 1950, when these initial calculations were instigated, the Hubble constant was believed to hold a value leading to the age of the universe being only of the order of 10^9 years. Since this figure was so much less than the time apparently required for the mass fraction of helium to be explainable from astrophysical processes, it was decided that it needed to be explained via primordial synthesis in the very early universe. The first crucial realisation to follow this decision was that it could be true only if the energy density of radiation in the early universe was large compared with the rest mass energy of matter. Accepting this was a major change in thinking for many since, up to that point, the opposite had been assumed true. An immediate consequence was that the radiation temperature had to be inversely proportional to the square root of the time. Up to this point, the argument was not unreasonable given the initial assumptions but what followed was a completely ad hoc step and it should be noted that it remains *ad hoc* today. The mass density of stable non-relativistic particles – neutrons and protons – decreases with the expansion of the universe and Alpher and Herman denoted this by ρ and took

$$\rho = 1.7 \times 10^{-2} t^{-3/2} \text{ g/cm}^3.$$

Here it is the choice of the coefficient of proportionality as 1.7×10^{-2} which is the *ad hoc* step. There is absolutely nothing in the theory of the big bang which actually fixes the value of this coefficient. It is a choice made quite freely but a choice which has the enormous, but to many acceptable, effect of ensuring that the calculated value for the mass fraction of helium is indeed 0.25, in accordance with the observed value. This must mean, however, that as the value of the said mass fraction changes, as it surely must over an extended period of time, the value of this constant of proportionality must change also. The obvious question to follow then is, does the big bang theory, therefore, actually *predict* the correct value for the mass fraction of helium? The answer has to be an emphatic 'No'!

The Common Perception.

It is, unfortunately, true to note that often, at the end of their undergraduate days, many students emerge totally convinced that the big bang theory correctly describes the beginnings of our universe and also many of its subsequently developed properties. They believe it to be the only theory which explains the cosmic microwave background radiation; they believe it to be the only theory to explain the mass fraction of helium. This, and much more, has all been learnt in undergraduate courses as being absolutely sacrosanct. Further, these beliefs are vigorously supported by so many popular science books, such as Simon Singh's *Big Bang*, and by many popular science lectures. Young people with impressionable minds leave such talks totally convinced that they have just been exposed to an enunciation of the complete truth regarding the birth of our universe. But have they? They will have been told, amongst other things, that the cosmic background radiation was discovered by Penzias and Wilson in 1965. McKellar's work will have been ignored. The steady state theory will have been dismissed totally with hardly a glance in its direction and no mention will have been made of the newer modified theory. The constant need to add to, and modify, the original Big Bang theory with entities such as dark matter and dark energy will have been glossed over completely. Herein lies a very real danger. The scientists of tomorrow are not being trained to have open questioning minds. Rather they are having their minds programmed to be closed to all thoughts which might possibly conflict with 'conventional wisdom'. The message often appears to be delivered with what amounts to an almost religious fervour, – what might be termed scientific evangelism.

It must be remembered that the steady state theory is still summarily dismissed as a serious attempt to explain the universe in which we exist. However, at this point in time, it should be noted that, even without the latest modifications to the theory, the advocates of steady state had answered many of the criticisms of that theory quite convincingly. The whole history of what Hoyle and his associates term 'the war of the source counts' provides a classic example of this. The details of this controversy are well documented, by those deeply involved on one side of the argument, in *A Different Approach to Cosmology* by Hoyle, Burbidge and Narlikar [10]. Here it is discussed in detail how, initially, it appeared that the steady state theory indicated incorrect results when it came to examining radio sources and their distribution. Essentially, it seemed that the data collected allowed either of two possible conclusions to be drawn. Ryle and his collaborators at Cambridge took one view; Hoyle subscribed to the alternative. This meant that Ryle and his supporters viewed the data in a way which opposed the validity of the steady state theory. The argument certainly raged fast and furious for many years but, in the end, following queries raised by Robert Hanbury-Brown at the Paris Symposium as early as 1958, the truth finally emerged following work published in 1988. In truth, some objections to the original steady state theory were destroyed at this point. However, this occurred some thirty years after the queries first erupted onto the scientific scene. Too much time had elapsed; too many opinions had been irrevocably formed; there was little or no chance that any change in popular scientific opinion would be accomplished. The modified theory, presented so eloquently in the above-mentioned book, is also not likely to create a revolution in scientific thought on this matter, - at least not immediately. Positions are far too entrenched; too much 'face' – and, possibly more importantly, too many positions of power and influence – would be lost if any senior scientist completed a *volte-face* on this issue. It is also sad to realise that many have been deterred from studying the steady state theory because it is felt by so many to have been disproved by observations and, therefore, merits no further study. On the face of it, this is a not unreasonable stand-point, but no-one can claim seriously that there is a single undisputed theory describing all aspects of our universe and its origins. True the big bang theory seems, in some ways, the most successful theory so far but, at best, that is all that it is, - the most successful theory so far. In all aspects of science, practitioners should remain open-minded and, in this particular area, more so probably than in others. It is incumbent on all – amateur as well as professional – to keep all options open and that means remaining fully up-to-date and conversant with all of the modified steady state theory, as well as the present version of the big bang.

Some Alternative Ideas.

Another problem associated with the Big Bang concerns the apparent lack of antimatter in the present Universe. The question of whether or not there is actual predominance of matter over antimatter is not necessarily a trivial one. In the middle of the last century, Hannes Alfvén and Oskar Klein suggested cosmological models which start with perfect symmetry between matter and antimatter. Subsequently in the theories these two components which comprise the Universe separate into matter-dominated and antimatter-dominated regions. Several objections were raised concerning this theory but an important one involved the manner of separation of the regions of matter and antimatter, since it was understood that even intergalactic space contains a small amount of matter and so galaxies could not be completely separate from antigalaxies. Alfvén [11] did propose a possible mechanism for achieving the required separation but most astrophysicists remained sceptical.

The mechanism proposed by Alfvén was effectively a generalisation of a phenomenon investigated in the 19th century by a German physician, Johann Leidenfrost. It was noted that, if a drop of water is placed on a surface whose temperature is in the region of $100^{\circ}C$, it will evaporate almost immediately. However, if the surface temperature is several hundred degrees, the drop does not boil off immediately; rather it becomes smaller gradually before disappearing completely. The explanation is that, at the higher temperature, as the drop evaporates, a layer of steam forms between the drop and the surface and this layer acts to insulate the drop from the surface so that heat is conveyed from the surface to the drop more slowly. Alfvén's idea was that a similar situation might exist in some circumstances between matter and antimatter.

Another model introduced just a little later in the 1970's by Omnès, Stecker and others had as an initial state a mixture of matter and antimatter separated by a Jordan surface, which is a simple closed curve separating two different components, each of which is fully connected. This state was referred to as an 'emulsion'. However, before too long, these efforts were abandoned because it emerged that separation on the scale of clusters of galaxies was needed to satisfy the then current observations but the model was found unable to demonstrate that coalescence could continue long enough for the accumulation of matter and of antimatter to grow even to the size of galaxies, let alone clusters of galaxies, before separation occurred. The problem of an initial baryon, anti-baryon asymmetry, necessary in today's dominant model to ensure the apparent dominance of matter in the Universe as it is today, remains. The fact is that the existence of an initial imbalance between baryons and anti-baryons is a purely ad hoc assumption. That being so, people have continued to speculate on the presence of antimatter in our Universe, even though the models of Alfvén, Omnès and others have long since been discarded. However, it is possibly of interest to note that, although, as mentioned, Omnès and his co-workers referred to a state as an 'emulsion', at no time did they utilise the properties exhibited by an actual emulsion in their deliberations. It is worth noting these particular properties and contemplating the effects of incorporating them into the model.

An emulsion is a mixture of two substances which normally wouldn't mix; that is, a mixture of two immiscible substances. One, referred to as the dispersed phase, is dispersed throughout the other, referred to as the continuous phase. Again, emulsions fall into two categories; colloidal emulsions which are stable so that one phase will remain dispersed throughout the other over a period of time, and non-colloidal emulsions which are unstable and in which the two components tend to separate out. On occasions, substances known as emulsifiers may be added to stabilise an emulsion. A very typical example of an unstable emulsion is provided by salad dressing. In this example, as is well known to all, the emulsion will separate out very quickly unless shaken very vigorously. However, for present purposes, this common example is worth bearing in mind as it is an example of an emulsion which illustrates very clearly what an emulsion is, how it looks and how it behaves.

In the original Omnès model, although the term 'emulsion' was used, the situation envisaged was more a mixture of individual blobs of matter and antimatter; there seemed no notion of one phase being dispersed throughout a second phase which remains fully connected. Normally, the two substances forming an emulsion will separate out over time if left undisturbed but the situation in the early universe described by Omnès was certainly not undisturbed, more akin in fact to the situation of a violently shaken salad dressing. However, simply introducing the notion of a genuine emulsion into the discussion cannot, of itself, help in the resolution of the problem of the missing antimatter since no conglomerations of antimatter have been identified in the Universe. Recently, an ingenious suggestion [12] has been advanced in an attempt to rectify this and that suggestion is that what might be termed the cores of black holes are all, both primordial and supermassive black holes, composed of antimatter. With the popular modern notion of a black hole, such a suggestion would mean all the antimatter being hidden from view inside the event horizon of the black hole. Also, considering the sizes of the postulated supermassive black holes, it is relatively easy to see how an equivalence of content of matter and antimatter in the Universe could be achieved; indeed, in the above mentioned article[12] some rough figures are included to support the plausibility of this assertion.

However, what if matter manages to cross the event horizon and come into contact with the antimatter? Obviously, any matter/antimatter contact will result in the annihilation of both but, in the model, the annihilation rate would be slowed down tremendously due to the antimatter being condensed into an extremely small body. Also, this annihilation would occur inside the event horizon and so there need not be any observation of resulting radiation. Further, it is suggested that such annihilation might not proceed too rapidly if a Leidenfrost layer, such as suggested by Alfvén, were to exist inside the event horizon. One further point occasioned by this idea is that such matter/antimatter annihilation could help the gradual evaporation of the black hole without recourse to the possible phenomenon of Hawking radiation, if such evaporation does, in fact, occur as speculated.

In the discussion so far, the role of the event horizon has been simply to prevent evidence of any possible matter/antimatter annihilation being viewed by observers; apart from that possibility, it appears to play no significant part in the model. Event horizons, though, are only part of the notion of a black hole which seems to emerge from the theory of general relativity. In the simplest case of an uncharged, non-rotating black hole, the starting point for discussion of the model is the Schwarzschild solution to the Einstein field equations but, as has been pointed out on numerous occasions [13], the popular version of that solution on which this deduction is based is not actually Schwarzschild's original solution, as is easily verified by referring to his original article and comparing it with the popular version which appears in so many textbooks. Schwarzschild's original solution does not possess the singularity which leads to the idea of a black hole. Hence, serious question marks hang over the modern notion of a black hole, added to which, as again has been pointed out on numerous occasions[13], so far no black hole candidate has satisfied the fundamental inequality to be satisfied by the ratio of its mass to its radius; that is, the inequality

$$M/R \geq c^2/2G = 6.7 \times 10^{26} \text{kg/m}.$$

However, even if the modern notion of a black hole has problems, theoretically the idea put forward by Michell in 1784 [14] and based on purely Newtonian principles is sound. Michell investigated the problem of a body with an escape speed greater than, or equal to, the speed of light. He found that the mass and radius of such a body would satisfy the same inequality as that mentioned above for a black hole as derived from the principles of general relativity. Since the event horizon plays so small a part in the above mentioned model of a balanced matter/antimatter Universe, it would not seem too much of a problem to substitute a Michell dark body instead of a black hole in that model. The term 'dark body' is used more correctly to describe the Michell idea since, as was pointed out by McVittie [15], if such a body exists, it would be simply a very dense body which could be approached and, in fact, viewed from a suitable distance, unlike the modern notion of a black hole. Obviously, this latter comment is in accordance with the usual meaning of a so-called 'escape speed'. It follows that the ideas advanced in the mentioned recent article [15] would hold if the bodies referred to were Michell type dark bodies of the appropriate size rather than conventional black holes since, although such objects wouldn't be hidden behind an event horizon, they would be effectively hidden from view by the very fact that even light would be unable to escape completely from them. Also, as with the suggestion based on black holes advocated in reference 12, any annihilation occurring would be slowed down to a great degree by the antimatter being condensed into an extremely small compact body. Of course, with no event horizon, if the dark body was composed of antimatter, any annihilation with nearby matter could only be prevented, or the effects slowed down, by the Leidenfrost layer solution as advocated originally by Alfvén. That in itself is no drawback to this modified suggestion since it is such a Leidenfrost layer which proves so important in the model suggested. It might be commented also that, in the case of a Michell dark body, the visibility referred to above would not mean that photons would reveal the presence of annihilation reactions since such photons would be degraded in energy and would not be what would be expected from annihilation. Of course, all of this particular discussion of the matter/antimatter problem is basically dependent on the big bang model being accepted as fundamentally correct. If it is not, then no immediate argument springs to mind to suggest the existence of antimatter in the Universe, at least not in quantities comparable with the amount of matter actually observed. Of course, consideration of this suggested model for the possible existence of comparable quantities of matter and antimatter in the Universe offers yet another possibility for examining the validity of the big bang model. As always, it should be remembered that the big bang is simply a theoretical model of how the Universe originated and developed and, as such, it must be open to observational and experimental checks in an attempt to establish how accurate a model it is or, in fact, if it is valid at all.

However, to return to the actual big bang theory, a further problem faced by the adherents to the theory is the seemingly constant need to add to the basic theory in order to overcome problems. Obvious examples of this are the introduction of the ideas of inflation, dark matter and even dark energy. It is, however, the first of these additions to which attention must be turned. The big bang model was faced with the 'horizon' and 'flatness' problems. The first of these arises from the prediction that the Universe is both homogeneous and isotropic, which implies that, in the early Universe, disconnected regions would have had to have been in nearly the same state to achieve the present-day homogeneity. The lack of contact makes such a scenario extremely unlikely. The second paradox concerns the extrapolation of the present value of the ratio of the energy density of the Universe to the critical energy density back to the big bang. When this is done, the extremely unlikely value of nearly unity is found. In 1981, Guth [16] attempted to address these by releasing the assumption of the adiabaticity of the early expansion of the Universe. This resulted in the so-called inflationary scenario, which supposes that a supercooling of the material of the Universe led to a period of exponential growth involving the release of the latent heat of the phase transition and an increase in the entropy of the Universe. Modifications to this basic model were introduced by Linde [17] and Hawking and Moss [18] to attempt to overcome the fact that it would produce large inhomogeneities which are incompatible with observation. The exponential dependence of the scale factor on the time is certainly a solution of the equations of general relativity, but the association of the release of a latent heat is not. This central objection went unnoticed until relatively recently [19].

An expanding Universe, as suggested by Hubble's observation of galactic expansion, will involve progressively increasing compression in the past. All that the inflation hypothesis was designed to do would be achieved by a speed of light which increases with increasing temperature. Of course, this alternative description of the past is not compatible with the universal application of the principle of general relativity which requires a universal speed for light.

It is not without interest to realise that additions to the big bang theory are accepted unerringly. Seemingly, no questions are raised when these new notions such as inflation, dark matter and dark energy are introduced in attempts to preserve this theory as the only acceptable explanation for our universe as we see it. However, there doesn't appear to have been any significant upsurge in interest in the steady state theory since the publication of all the material – both strictly academic and semi-popular – advocating modifications to that theory. Many will claim this due to the fact that the theory is quite simply incorrect, but the facts don't support this view. Neither do they support the view that the big bang theory is true beyond all reasonable doubt. In reality, the truth must lie either somewhere between these two extremes or possibly completely outside these two interesting attempts buried in some, as yet, totally unknown theory. We really truly understand very little, however great mankind's scientific achievements may be thought to be. When we understand in detail what is meant by terms such as 'force' and 'mass', then we will be on the way to a complete understanding of our universe and all that exists in it but, until that time, it seems sensible to retain all options and that must include both the big bang model and the steady state theory, together with any other thoughts, as possible explanations. Prominent among these other thoughts must be the so-called 'tired light' theory. So much in our presently accepted theories depends on the interpretation of the red-shift phenomenon. It is commonly accepted, as has been mentioned already, that this red-shift is brought about by the Doppler shift of light due to the recession of distant galaxies. However, at least theoretically, other explanations are feasible. A brief outline of the worries expressed by Halton Arp has been discussed earlier. However, another possible explanation for the existence of the observed red-shifts is provided by the notion of 'tired light'. Here the basic idea is that quanta of light could actually lose energy during their journey through space from distant galaxies to us. The suggested decrease in photon energy would result in an increase in wavelength that would be proportional to the distance travelled. This would, of course, be viewed as a reddening. Another contributory factor to this reddening of light could be provided by scattering by particles of intergalactic dust. Probably the effect of scattering by dust particles may be discounted at this stage, though not entirely forgotten, because such scattering would be expected to result in a broadening of the spectral lines and that is not observed. However, the general notion of 'tired light', while dismissed almost out of hand by most workers in the field, cannot be totally abandoned as yet. Firstly, it is a theory which has a long history and which has never gone away completely. It has been advanced and supported by a powerful array of physicists from Max Born to Jean-Pierre Vigier. This, in itself, is not sufficient to make the theory acceptable, but it is surely a good enough reason for it to be taken seriously. Some wish to dismiss it on the grounds that only in Big Bang cosmology is there a satisfactory explanation provided for the origin of the cosmic background radiation and for the abundance of the light elements. However, as has been seen already, this is simply not true. The case of the steady state theory proves this beyond reasonable doubt. Whether one believes or disbelieves the steady state theory or, for that matter, the big bang theory, it is certainly true to say that, in attempting to destroy the steady state theory, the truth was not to the fore. It is disturbing to realise that this explanation is the one advanced for dismissing so many suggestions and it is no more true today than it was when first put forward and agreed. 'Tired light' may not be a true explanation for any of the questions arising in cosmology but, like anything else, it deserves to be viewed with a completely open mind before a decision is reached. Once again it is seen that this is the true problem facing cosmology as a whole and the big bang theory in particular – both must be viewed and assessed with completely open minds. Personal preferences and prejudices have no place in the evaluation of a scientific theory. The task must be accomplished purely by using the accepted methods of science and known scientific knowledge - always realising, of course, that any conclusion will be subject to limitations placed on its validity by the extent of such knowledge at any one time.

Yet another major problem facing this area is associated with the advance of knowledge. In this colossal area, knowledge advances through careful, painstaking observation of the cosmos. All the observations made must then be processed most carefully. This again is something which is not quite so straightforward as might appear at first. Quite frequently, data has to be analysed statistically and it is crucial that this is done completely honestly. There must never ever be even a suspicion that an effect is claimed which might be simply due to the statistical package used for the analysis. Hence, this again is something which must be undertaken by truly open-minded people and making use of professional statisticians to analyse data – rather than it being done by those who might be thought to have a vested interest in the end result – could be a sensible way forward in this area. Too often the impression is left that the conclusion announced is merely confirmation of the result 'expected' before the experiment or observation was begun. In a way, this brings a return to the case of Halton Arp. As has been noted earlier, many astronomers are said to doubt Arp's interpretation of the photographs he has taken and usually their scepticism is said to be based on some aspect of the statistical analysis of his data. It has been claimed, though, that if a continuous change in red-shift values could be measured along an apparently material link between a low red-shift galaxy and a high red-shift quasar, then Arp's view would be vindicated. However, it seems that no such effect has been found as yet, although strenuous efforts are said to have been made to establish the presence, or absence, of such an effect. This again raises the question of whether or not observers are finding what they want to find rather than the truth. Some ask at this point, 'What is truth?' No doubt a deep philosophical discussion could ensue here. However, suffice it to note that the Oxford Dictionary states that one meaning of the word 'true' is "in accordance with fact or reality, not false or erroneous". It goes on to state that 'truth' is the "quality, state, of being true". These elementary definitions of the two words give a clear everyday meaning of what they mean in the present circumstances. Indulging in philosophical discussions surrounding the meanings of words doesn't necessarily help anyone; it frequently serves simply to divert attention from the question at issue, - in this case that of the major problems facing science today. As with so many of the major controversies in science, positions have become entrenched, 'conventional wisdom' has become almost indelibly etched into the folklore surrounding the subject. Young scientists are, all too often, taught established truth as if it were religious dogma. They are not trained to really think; only to think along well-established lines – lines drawn by the 'Gods' of 'conventional wisdom'. This probably sounds harsh and seemingly linking science with religion again will undoubtedly offend some who feel the two separated by an infinite chasm. Unfortunately, the truth often does hurt and, in reality, young scientists are all too often indoctrinated with supposed 'facts', rather than educated to have open, enquiring minds. If the result of raising these unpleasant aspects of present day world science is to reintroduce an open questioning attitude into science, then the imagined hurt will have been more than worthwhile.

As an addendum to this discussion of the big bang model, it might be noted that an entire edition of the well-known and well-respected British Broadcasting Corporation's television science programme, Horizon, was devoted to the present-day search for dark matter [20]. The programme title was *Most of our Universe is Missing*; an eye-catching title guaranteed to attract viewers. It pointed out that some scientists feel it not known from what much of our universe is made; others argue that some presently accepted theories, such as Newton's law of gravitation, may be wrong – or, at least, only apply locally rather than globally. The programme itself contained much of genuine, but not probably general, interest. However, one worrying aspect in the present context was the fervour exhibited by several contributors in support of the big bang as explaining the origins of the universe. Only one really drew back to point out that the big bang is a theory, and only a theory! As was asked in a recent letter to *The Observatory* [21], "When will the *Cosmological Establishment* stop calling their theory the truth, the whole truth, and nothing but the truth?" Anyone who questions it is said to belong to a minority. Apparently, most cosmologists would offer strong odds on there having been a 'big bang', feeling that "everything in our observable universe started as a compressed fireball, far hotter than the centre of the Sun". The idea that this scenario is questioned by a minority only would seem true, but largely because so many in science possibly feel it in their own best personal interests to conform to the imposed dictates of 'conventional wisdom'. Of course, in these terms, that 'minority' might really be a 'silent majority'. As for those outside professional scientific circles, those who in the final analysis pay the bills, they have been subject to so much publicity, via all media forms, in favour of this theory to the exclusion of all else, that it is no wonder they believe it to be an unassailable truth, not simply a mere theory. However, as another contributor to *The Observatory* pointed out [22] because the steady state theory appears to provide precise predictions, it seems to have suffered in comparison with other theories, such as the Big Bang, which allow scope for empirical adjustment. This writer felt it precisely this which made the steady state theory a good theory and seemed to feel it likely that that theory would return eventually in some form. Be that as it may, it is undoubtedly of interest to speculate on what the future holds in this field, but one thing is absolutely certain, for real progress to be made, investigators must retain open minds; very little should be totally discarded at this juncture. In the present atmosphere that seems a lot to ask, but it is absolutely essential if science is to advance positively!

Plasma Cosmology/Electric Universe.

Of course, all of the discussion up to this point has been based on a theory dependent solely on possible effects due to the force of gravity. It might be noted that, following the introduction of Newton's mechanical ideas, work still proceeded apace investigating electromagnetic phenomena and this continued at least into the earlier years of the twentieth century, as is evidenced by the contents of J. J. Thomson's book *Electricity and Matter* [23] (Archibald Constable & Co. Ltd, Westminster, 1904). However, this book provides but one example to illustrate the very real emphasis on work involving the effects of the electric and magnetic fields. However, after those early years of the century, the emphasis seems to have shifted to explanations of phenomena purely in terms of gravitational effects as far as most mainline research was concerned. Considering that it is accepted that much of the matter in the Universe is in the form of plasma, this seems a retrograde step and this view is surely strengthened when the work of such as Kristian Birkeland and Hannes Alfvén is considered. One may only speculate as to why the emphasis of much scientific research changed in this way. However, thanks to people like Birkeland, Alfvén and (more recently) Peratt, work in the areas of electromagnetism and plasma physics did continue and it should be noted that much of the work on plasmas has been via laboratory experiments, so hard experimental evidence is available to support any claims made.

The work on plasmas and other electromagnetic phenomena has inspired the examination of astronomical phenomena in these terms and has resulted in the so-called Electric Universe idea as expounded, for example, in the books *The Electric Universe* by Wallace Thornhill and David Talbott [24] and *The Electric Sky* by Donald Scott [25]. Reading through this material makes one aware that, while like orthodox accepted theory, the electric universe ideas are supported by much computer modelling, it can also draw on parallels in astronomy with plasma phenomena observed in the laboratory. Admittedly, drawing such parallels involves scaling up tremendously but assuming this possible is little different from assuming that laws seemingly applicable here on the Earth are also applicable in the Solar System and, indeed, throughout the Universe. However, at least visually, some of the phenomena observed in the laboratory are very like what is observed by some of the most powerful of telescopes; - electric currents in plasma naturally form filaments due to the so-called 'pinch effect' of the induced magnetic field. Electromagnetic interactions cause these filaments to rotate about one another to form a helical 'Birkeland Current' filament pair and this is very much the structure seen in the Double Helix nebula near the galactic centre; again, the Hubble image of the planetary nebula NGC6751 looks remarkably like the view down the barrel of a plasma focus device. Examples such as these prove nothing but should awaken people to the possibility of alternative explanations for astronomical phenomena.

Much of the laboratory work originated with the work of Kristian Birkeland more than one hundred years ago. It was during his Arctic expeditions at the end of the 19th century that the first magnetic field measurements were made of the Earth's polar regions. His findings also indicated the likelihood that the auroras were produced by charged particles originating in the Sun and guided by the Earth's magnetic field. Birkeland, though, was an experimentalist and is still known for his Terrella experiments carried out in a near vacuum and in which he used a magnetised metallic sphere to represent the Sun or a planet and subjected it to electrical discharges. By this means, he was able to produce scaled down auroral-type displays as well as analogues of other astronomical phenomena. These claims, however, were only vindicated finally by satellite measurements in the 1960's and 70's. To that point in time, his experimental and observational achievements had tended to be overshadowed by the purely theoretical predictions and explanations of the geophysicist, Sydney Chapman. Once again, powerful mathematics seems to have held sway over the more expected techniques of physics – experimentation and observation, with mathematics a mere tool to be used when necessary. This is not to decry Chapman's work but to emphasise the overwhelming importance of the physics when investigating natural phenomena.

Birkeland also showed experimentally that electric currents tend to flow along filaments shaped by current induced magnetic fields. Of course, this confirmed observations of Ampère that indicated that two parallel currents flowing in wires experience a long range attractive magnetic force that brings them closer together. However, as plasma currents come closer together, they are free to rotate about each other. Such action generates a short range repulsive magnetic force which keeps the filaments separated so that they are, in effect, insulated from each other and able to maintain their separate identities. The end effect is for them to appear like a twisted rope and it is this configuration which is termed a 'Birkeland current', as was mentioned earlier when the Double Helix nebula was noted as a possible example. Satellites orbiting above the auroras in the 60's and 70's were able to detect a movement of ions, indicating that electric currents were present. Later missions found quasi-steady electric fields above the auroras following the magnetic field lines, thus lending some credence to Birkeland's claim of the existence of an electric circuit between the earth and the Sun. Some may be sceptical of this latter interpretation but it is undoubtedly true that much of the material in the Universe is in the form of plasma and there is certainly electric and magnetic activity occurring in abundance. This means there are numerous very good reasons for considering the effects of the electromagnetic force in the Universe, one of which could be the resolution of the problem of the missing mass.

However, precisely what is the Electric Universe? In truth, it is really simply an hypothesis, a new way of interpreting known data by utilising both new and well-established knowledge relating to electricity and plasma. It should be emphasised immediately that, in this new interpretation, gravity still has a role to play but it is a secondary one since the electric force is so much more powerful. A major point to be stressed from the outset is that, in this interpretation of astronomical phenomena, scientists are able to call on evidence from laboratory based experiments to help form and support suggested explanations for a wide variety of phenomena. It has been found that, as explained in more detail in the above-mentioned books, an electrified plasma in a laboratory is a good model for providing possible explanations for many recently observed astronomical phenomena which, in several cases, have caused puzzlement for astronomers seeking explanations via more orthodox gravitationally based theories. This is not to say that gravity is ignored and regarded as irrelevant; rather, the possible effects of the electromagnetic force on astronomical phenomena are investigated while still recognising the importance of gravitational effects. In the electric universe, the gravitational systems of galaxies, stars, moons and planets are felt to have their origins in the proven ability of electricity to generate both structure and rotation in plasma. It is felt further that the force of gravity assumes importance only as the electromagnetic forces approach equilibrium. As has been noted already, great consternation has been caused in astronomical circles by the realisation that gravity, as presently understood, cannot explain much that is observed if the amount of mass available is as now felt to be present. Hence, instead of positing the existence of 'dark matter' or following the path of modifying Newton's well-tried law of gravitation, it is suggested here that the possible effects of the electromagnetic force be examined to see if, in conjunction with orthodox ideas on gravity, these puzzling observations can be explained.

A point which is often relegated to the background when discussing the solution of problems through the introduction of dark matter is the fact that the missing mass, if there really is any missing mass, is not absent homogeneously throughout the Universe; it is missing only in specific places - for example, in the outer regions of galaxies. Hence, possible solutions, such as the idea that neutrinos possess mass, which are essentially homogeneous in nature cannot be acceptable. It should be mentioned at this point though that, in the Electric Universe model, neutrinos do possess mass and are extremely important. They respond only weakly to massive objects such as stars and galaxies but form an extended atmosphere which, for example, refracts light around the Sun from distant stars and this offers an alternative explanation for the so-called gravitational bending of light. On the other hand, in this model, neutrinos are not required to explain galactic rotation although they must contribute to the masses of both stars and galaxies. Again, having some mass, neutrinos will not be distributed homogeneously.

However, returning to the realisation that much of the matter permeating the Universe is in the form of plasma, it might be remembered that these clouds of plasma respond to the well-known laws of Maxwell. Also, as pointed out by Scott in his book, another law, formulated by Lorentz, does help explain the galactic speeds alluded to earlier. This law states that *a moving charged particle's momentum (speed or direction) can be changed by application of either an electric field or a magnetic field or both.*

This seems a highly likely contributory factor, at least, causing galaxies to rotate as they are perceived to do but would indicate, contrary to the accepted view, that gravity has less to do with things than has been thought. However, it should be emphasised that nowhere is it being suggested that Newton's law of gravitation is wrong or in need of modification; it is simply being suggested that, in deep space where everything swims in a sea of plasma, the Maxwell – Lorentz electromagnetic forces dominate over those of gravity.

It might be remembered also that the Lorentz force alluded to here changes a charged particle's momentum and that change is directly proportional to the strength of the magnetic field through which the particle is moving. Further, the strength of a magnetic field produced by an electric current is inversely proportional to the distance from the current but the gravitational force between stars is inversely proportional to the *square* of the distance. This well-known difference between the two forces could lie at the heart of the problem of the galactic rotation curves; certainly it seems an avenue worth exploring further, especially considering the fact that more and more space missions are indicating that electromagnetic forces are distributed more widely throughout space and are, of course, many orders of magnitude stronger than gravitational forces.

Much time, effort and money is spent worldwide on producing elaborate computer programs which purport to support the prevailing belief in the big bang model as being the correct theory explaining how the Universe originated. However, as well as a great many laboratory experiments being performed to help establish plasma properties, it has been shown also, using the Maxwell and Lorentz equations, that streams of charged particles, such as are found in the intergalactic plasma, will evolve into the familiar galactic shapes under the influence of electromagnetic forces. The results fit extremely well with the observed velocity profiles in the galaxies and all this with no recourse to missing mass. Much of this simulation work has been carried out by Anthony Peratt and is reported in various issues of the IEEE Transactions on Plasma Science, a highly prestigious journal. It does seem, therefore, that the case for the existence of dark matter is questionable and hence, yet another query is raised concerning the validity of the big bang model.

Conclusion.

Possibly that last word – 'model' – is the most significant one used for that is *all* the big bang is, a model! Admittedly, it has been successful to some extent in attempting to explain the origin and subsequent development of our Universe, but much of that success has really been apparent due to the fact that so much information and so many alternatives have been kept hidden from public scrutiny. Here the ideas of the so-called Electric Universe have been used mainly in relation to an examination of the question of the existence, or not, of so-called dark matter, that mystical entity introduced to shore up the wobbling framework of the big bang model. However, it has been shown that this is another means of attempting to explain at least some aspects of the behaviour of our Universe. It seems immediately apparent though that this theory simply *must* play at least some part in the explanation since the electromagnetic force is so much more powerful than that of gravity. Again, so many more modern space missions are indicating more and more effects of both electric and magnetic fields in the space above us that it cannot be long before it is acknowledged that the effects of these can be ignored no longer.

There is little doubt that, just as he has done throughout the ages, man will continue to search for as much knowledge as possible about the Universe in which we live. Again there is little doubt that at least part of that search will involve seeking the answer to the age-old question 'How did it all begin?' Whether or not man's intellect is capable of discovering the true answer to this question is another matter; he will continue to search. What really matters is that that search should be carried out scientifically; workers must be scrupulously honest in all their work and in the reporting of that work whether in academic journals or in more popular publications designed to keep the public, which eventually pays for all this endeavour, informed of progress. The cult of 'conventional wisdom' must, therefore, be eradicated and scientific research must be conducted in a completely open minded manner with no single theory or model being allowed to dominate purely to preserve the status of powerful individuals. As Hannes Alfvén said in his Nobel Lecture of December 1970, "The centre of gravity of the physical sciences is always moving. Every new discovery displaces the interest and the emphasis." Maybe those working in the fields of astronomy/astrophysics and cosmology especially should take note of these words of wisdom uttered by an acknowledged scientific thinker and open their minds to other possibilities when attempting to solve problems in their preferred scientific domains.

References.

1. G.Lemaître, 1927, Ann. De la Societe Scientifique de Bruxelles, 47, 49

2. G.Gamow, 1946, Phys. Rev., 70, 572

3. A. Penzias & R. Wilson, 1965, Ap. J., 142, 419

4. F. Hoyle, G. Burbidge & J. V. Narlikar, 2000, A Different Approach to Cosmology, (Cambridge U. P., Cambridge)

5. H. Bondi & T. Gold, 1948, M. N. R. A. S., 108, 252. F. Hoyle, 1948, M. N. R. A. S., 108, 372

6. A. Eddington, 1988, The Internal Constitution of the Stars, Cambridge U. P., Cambridge.

7. G. Gamow, 1946, Phys. Rev., 70, 572

8. G. Gamow, 1948, Phys. Rev., 74, 505. R. A. Alpher & Herman, R., Phys. Rev., 75, 1084

9. F. Hoyle, 1946, M.N.R.A.S., 106, 343

10. F. Hoyle, G. Burbidge & J. V. Narlikar, 2000, A Different Approach to Cosmology, (Cambridge U. P., Cambridge)

11. H. Alfvén, 1966, *Worlds-Antiworlds*, Freeman & Co., San Francisco.

12. H. D. May, 2010, *Another Look at the Cosmological Model of Omnès*, viXra:1011.0017

13. J. Dunning-Davies, 2007, *Exploding a Myth*, Horwood Publishing Ltd., Chichester.

14. J. Michell, 1784, Philos. Trans. R. Soc. **74**, 35.

15. G. C. McVittie, 1978, The Observatory, **98**, 272.

16. A. Guth, 1981, Phys. Rev. D, 23, 347

17. A. D. Linde, 1982, Phys. Lett. B, 108, 389

18. S. W. Hawking & I. G. Moss, 1982, Phys. Lett. B, 110, 35

19. B. H. Lavenda & J. Dunning-Davies, 1992, Found. Phys. Lett. 5, 191

20. Horizon, B.B.C. 9th February, 2006

21. A. Welch, 2006, *The Observatory*, **126** (no. 1190), 51

22. P. Fellgett, 2006, *The Observatory*, **126**, (no. 1190), 51

23. J. J. Thomson, 1904, *Electricity and Matter*, Archibald Constable & Co.Ltd., Westminster

24. W. Thornhill & D. Talbott, 2002, *The Electric Universe*, Mikamar Publishing, Oregon

25. D. E. Scott, 2006, *The Electric Sky*, Likamar Publishing, Oregon

7. The Enigma that is Light.

Jeremy Dunning-Davies,

&

Richard Lawrence Norman

Abstract.

The presently accepted notion of wave/particle duality, especially when applied to light, is undoubtedly a cause of great unease for many. Here the issue is examined afresh in the light of ideas put forward in recent years, especially by Mayants, while not forgetting the contributions of those great scientists of the past. Explanations and interpretations are then offered to account for seemingly paradoxical effects.

Introduction.

What is the true physical nature of light? This is a seemingly simple question which has been around in science for centuries. Newton favoured a particle theory but found difficulty when trying to explain the so-called Newton's rings with this theory; then, Young's experiments appeared to indicate a wave nature. The problem of attempting to reconcile the apparent wave and particle properties of light seemed for years to be an intractable one. However, reconciliation came in the wake of the quantum mechanical notion of wave/particle duality; in other words, light could display both wave and particle properties depending on the physical situation under consideration. Mathematically this might seem an acceptable resolution of this considerable problem but, physically, it seems it has always left people with at least a feeling on unease. In view of material, both experimental and theoretical, which has accrued over recent years, it is perhaps time to look again at this question with (hopefully) a completely open mind. The latter point is vitally important because some material might be considered to cause a 'rocking of the boat' in areas of science deemed sacrosanct by some.

Wave or particle?

Much of the more modern work carried out in addressing the meaning of wave/particle duality has been by Lazar Mayants and this is worth considering in some detail. It probably goes back at least to the appearance of his first book, *The Enigma of Probability and Physics*[1], continues through a number of publications, with one of 1989[2] being of particular relevance here, and culminates in the lucid overall discussion in his second book *Beyond the Quantum Paradox*[3]. To start, it seems worth considering in some detail some of the material contained in Mayants 1989 article cited under reference 2. Here he examines, for example, the phenomenon of particle diffraction but first examines straightforward diffraction which is known to occur when a series of waves of the same frequency encounters obstacles. The resulting diffraction pattern is determined by the geometry associated with the total system involved, together with the wave-length of the wave involved. For a real physical wave process, the detail will be determined eventually by the wave equation and the relevant boundary conditions. The relevant wave equation has the form:

$$\nabla^2\varphi = \left(1/v_p^2\right)\partial^2\varphi/\partial t^2, \qquad\qquad \text{(a)}$$

where v_p is the phase velocity of the waves and φ is a quantity whose magnitude squared determines the diffraction pattern.

However, what is the actual position concerning particle diffraction? If the conventional belief that individual particles possess inherent wave properties is true then any such particle should have some property, akin to φ, obeying the above wave equation. If such a property does exist then the diffraction pattern should remain unaltered, regardless of the intensity of the beam, but gradually weakening as the beam does. However, experimentation does not support this. If only a few particles are used, no continuous diffraction pattern appears; a few points on the display are all that do appear. In the case where a large number of particles is used, the picture seems to be a normal diffraction pattern but, in reality, it isn't; it simply consists of a very large number of points which appear to merge together to produce a familiar diffraction pattern. One might say that the perceived result is essentially a statistical one in the sense that a very large number of particles is involved and such numbers may only be treated effectively by statistical considerations. These remarks have been phrased to refer to particles – any particles – and, therefore, would refer to photons if photons are considered as particles.

In an actual particle diffraction experiment, a beam of *concrete* particles (to use Mayants terminology) is concerned and the experimenter considers the experimental statistical distribution of the coordinates of these diffracted *concrete* particles. However, in the theoretical situation, attention turns to the relevant probability distribution of the coordinates of what are, in effect, corresponding *abstract* diffracted particles. It is this rather subtle distinction between the *concrete* particles of the experimenter and the corresponding *abstract* particles of the theoretician which lies at the very heart of Mayants' argument. This seemingly obvious distinction between abstract and concrete objects is an error hiding in plain sight. To see the distinction with clarity, allows the removal of many apparent paradoxical contradictions. As Mayants says[3], "It is these two principle features of abstract objects—the nonexistence in reality and the lack of definite values of many properties—which differentiate them from the corresponding concrete objects." "Which comes first, the chicken or the egg" refers to an abstract chicken and an abstract egg. The question being based on an abstract object can not be answered, but that is not important, as the abstract object "the chicken" does not have particular properties or exist, rendering the question improper and trivial. Each real particular such bird exists in no temporal paradox but comes after the egg in which it was gestated, and before any egg it may itself produce. In like fashion, the subject of "a cat" in Schrodinger's paradoxical experiment, which it must be remembered was outlined in the original case to point up quantum theoretic inconsistencies at macro scale is again an abstract cat, in this case symbolizing an indeterminate probability distribution, which is itself again an abstraction with undefined qualities. The resultant paradox simply does not exist. Probability theory works, and an abstract set adequate to a concrete set gives good results in calculations, but in no case are abstract and concrete objects alike. Paradox itself is not paradox, but misunderstanding. *The world is made of concrete objects.* It is this which the theory in its end *result* must describe, and does.

As has been described in detail elsewhere[2], it follows that the probability distributions of physical quantities for an abstract physical system, which conform to real motion of the corresponding concrete physical system, are determined by the solutions to the Schrödinger equation

$$E\psi = H\psi.$$

For a free real particle, the Hamiltonian is given by

$$H = c(p^2 + m_0^2 c^2)^{1/2},$$

where, as usual, m_0 is rest mass and p is momentum. c is the speed of light in a vacuum. However, the operators for particle momenta are $p_a = -i\hbar\, \partial/\partial\alpha$ and $E = i\hbar\, \partial/\partial t$ Then the Schrödinger equation takes the form

$$c(-\hbar^2\nabla^2 + m_0^2 c^2)^{1/2}\psi = i\hbar\partial\psi/\partial t$$

which leads to

$$\hbar^2 c^2 \nabla^2 \psi = m_0^2 c^4 \psi + \hbar^2\, \partial^2\psi/\partial t^2 \qquad (b)$$

However, diffraction refers to a stationary state of the particle, determined by a specific value E of the energy which corresponds to a definite value of the momentum p and these are linked via

$$p^2 = (E^2 - m_0^2 c^4)/c^2.$$

It follows that

$$\hbar^2 \, \partial^2 \psi / \partial t^2 = -E^2 \psi \text{ and } \psi = -E^{-2} \hbar^2 \, \partial^2 \psi / \partial t^2.$$

Substituting in the first term of the right-hand side of (b) above gives

$$\nabla^2 \psi = c^{-2}(1 - m_0^2 c^4 / E^2) \, \partial^2 \psi / \partial t^2.$$

By putting $(E^2 - m_0^2 c^4)/c^2 E^2 \equiv 1/v_p^2$ in this equation leads to

$$\nabla^2 \psi = \left(1/v_p^2\right) \partial^2 \psi / \partial t^2,$$

that is, the well-known wave equation (a) with $v_p = E/p$.

At the very least, this would seem to indicate that particle diffraction is not a wave process but is, rather, linked to the probability distribution of particles in a stationary state corresponding with well-defined values of both energy and momentum for the particles. Again, it does suggest that the whole notion of wave particle duality should be re-examined with truly open minds. The above outlined theory is due, as stated previously, to Mayants – particularly in his cited article of 1989 - but is work which seems to have been forgotten by much of the scientific community and is certainly deserving of more public acknowledgement.

The Speed of Light.

What is really meant when people speak of the speed of light? What is meant when reference is made to the constancy of the speed of light? Popular interest concerning issues linked to the speed of light has probably increased since the popularisation of Einstein's theories of relativity. It is a popular misconception that Einstein's theory claims the speed of light to be a constant and that the theory leads to an ultimate speed for everything which is this constant speed of light. This, however, is only an incorrect public misconception.

It is important to remember that Einstein assumed the speed of light *in a vacuum* to be constant. Also, in several subsequent mathematical manipulations, the factor

$$(1 - v^2/c^2)^{-1/2}$$

appears, with *v* being the speed of the object under consideration and *c* the speed of light but, as emphasised above, the speed of light in a vacuum. It must always be remembered, though, that Einstein's theory was and is just that – a theory. Like any theory it will only hold when the assumptions made in constructing it hold; if any one of those assumptions ceases to be valid, it cannot be assumed the theory continues to be valid. This is, of course, true of any theory.

These points are important to remember since it is known, and has been known for a long time, that the speed of light is *not* constant; it certainly varies for light passing through different media. The speed of light passing through a medium of refractive index *n*, is *c/n*, where *c* is the speed of light in a vacuum. Hence, for light passing through a medium, such as water, which possesses a refractive index greater than unity, the speed of light will be substantially less than the value in a vacuum. Therefore, the ration v^2/c^2 in the factor mentioned above will be less than unity and so, no mathematical problems are encountered with this factor. However, there are media which appear to possess refractive indices less than unity and, in such cases, light will propagate at speeds in excess of the speed in a vacuum. This, in turn, raises questions about the above relativistic factor since if v is greater than c in this expression, mathematical problems do arise due to the appearance of a negative quantity whose square root is required.

What must be remembered here is that, as Santilli has explained it[4], special relativity was constructed to describe the propagation of light in a vacuum but not within physical media. Many of the results of special relativity have been validated on numerous occasions for point particles or electromagnetic waves moving in a vacuum but the theory is inapplicable for the movement of such in physical media because the speed of light is really a local variable dependent on the properties of the medium through which it is passing.

As with all physical theories, it is important to realise that they are just theories and, as such, are based on certain very definite assumptions. If any theory is applied in a situation where one or more of those basic assumptions is invalid, that theory cannot reasonably be expected to produce a satisfactory explanation of that situation. Note that this does not mean the theory is incorrect, it merely points out that it is invalid.

Axiomatic implications: Uncertainty, EPR, Popper, Bell and gravitation.

Uncertainty, as an inherent systemic property and the quantum uncertainty principle we attribute to Heisenberg, as well as the closely related wave/particle duality have been the target of much enquiry and by no means stand on certain and irrefutable ground, nor should they. The Einstein-Podolsky-Rosen (EPR) paradox, is in no way paradoxical. In fact, it reveals the uncertainty relation itself to be "paradoxical." Indeed, this simple thought experiment involving two particles moving along the same linear path in the same direction at the same speed, maintaining therefore fixed relative distance, does allow the precise simultaneous determination of both position and momentum of either particle. The thought experiment refers to concrete particles, and has a non-paradoxical outcome, where the uncertainty principle refers to quantum probabilistic calculations upon abstract objects, yielding a "paradox" when mistakenly applied directly to the particular, 'concrete' world.

Mayants is not the first to advance some of these ideas, which can be seen in the work of Popper in slightly different language[5]. The factual order of historical development points to an initial particle view of EM, with the field then later added as a secondary mathematical abstraction, which subsequently had the particle, the photon, emerge secondary to the field as an excitation[3]. Indeed, it appears we see the same confusion yet again, and perhaps it may be fruitful to restore the proper genesis of theory and realities, place the photon at the base of its collective wave propagation, and understand it is the source of any emergent field effects.

In reference 5 it is stated that:

"Max Born himself says about his statistical interpretation of wave mechanics: "The solution . . . was suggested by a remark of Einstein's about the connection between the wave theory of light and the photon hypothesis. The intensity [of course, what is meant is the square of the amplitude] of the light waves was to be a measure of the density of the photons or, more precisely, of the probability of photons being present."

"Thus, through Born's statistical interpretation of matter waves even the one problem of quantum theory which appeared not to be statistical - the problem of atomic stability - was reduced to, or replaced by, a statistical problem: Bohr's quantized "preferred orbits" turned out to be those for which the *probability* of an electron's being found on them differed from zero."

"All this is to support my thesis that the *problems of the new quantum theory were essentially of a statistical or probabilistic character.*"

However, Popper also draws this unusual, apparently contrary conclusion which will fit into place later:

"Thus the relativity to specification of which we have spoken is characteristic neither of quantum experiments nor even of statistical experiments: it is a permanent feature of all experimentation. (And a propensity relation might be regarded, and intuitively understood, as a generalization of a "causal" relation, however we may interpret "causality".) For this reason it seems to me mistaken to regard statistical laws, statistical distributions, and other statistical entities, as non-physical or unreal. Probability fields are physical, even though they depend on, or are relative to, specified experimental conditions."

In order to make sense of the above statement, it may be beneficial to take an elliptical pathway and consider the consequences of these insights as applied to one of the basic tenets supporting the current predominant quantum viewpoint: the Bell inequalities. Mayants' commonsense analysis will have come as an unwelcome surprise to some. However, any facts unearthed in a cogent analysis such as his must be accepted and it must be seen where they lead. It may be noted that Bell's inequalities suffer from the same logical error as the other 'paradoxical' constructs considered above: an erroneous substitution of abstract for concrete elements. Bell's inequalities are based on Bell's theorem, which is itself a derivative of Bohm's paradox, and Bohm's paradox confuses abstract quantum elements and concrete objects. It is argued[3], that the basic experiments upon which Bell's inequalities are based can apply only to large numbers of particle pairs and must represent a statistical expression, and so, it is therefore entirely *expected* that Bell's inequalities do not conform to experiments involving the real concrete system in question, as Bell's inequalities confuse abstract and concrete elements to assume simultaneous rotation amongst various axes in the case of *one individual particle*, which is physically impossible. Quantum physics in this light may be rightly seen as representing real non-paradoxical outcomes and Bell's inequalities are thereby revealed as flawed at their axiomatic basis, hence the apparent but nonexistent paradox.

From this new vantage Mayants[3] informs us of the *ordinary view* of the consequences implied which place Bell's ideas and nonlocal faster than light effects on one side of the scales, and on the opposing side of the balance we find realism and the common if incorrect assumption that *nothing moves faster than c*. Recall that Einstein's limit of c refers to propagation through a vacuum. Does Einstein's c hold good as a matter of consequence to defeat nonlocal theory, if Bell's ideas are not correct? What of nonlocality? What of light? Do physical processes move faster than light in a nonlocal way and, if so, which ones?

In Tom Van Flandern's essay[6], *The Speed of Gravity what the experiments say*, a solid and specific empirical answer is provided [see original article for embedded references]:

"The most amazing thing I was taught as a graduate student of celestial mechanics at Yale in the 1960s was that all gravitational interactions between bodies in all dynamical systems had to be taken as instantaneous.

. . . Yet, anyone with a computer and orbit computation or numerical integration software can verify the consequences of introducing a delay into gravitational interactions. The effect on computed orbits is usually disastrous because conservation of angular momentum is destroyed.

. . . While relativists have always been partial to the curved space-time explanation of gravity, it is not an essential feature of GR. Eddington (1920, p. 109) was already aware of the mostly equivalent "refracting medium" explanation for GR features, which retains Euclidean space and time in the same mathematical formalism. In essence, the bending of light, gravitational redshift, Mercury perihelion advance, and radar time delay can all be consequences of electromagnetic wave motion through an underlying refracting medium that is made denser in proportion to the nearness of a source of gravity. (Van Flandern, 1993, pp. 62-67 and Van Flandern, 1994) . . . The principal objection to this conceptually simpler refraction interpretation of GR is that a faster-than-light propagation speed for gravity itself is required. In the context of this paper, that cannot be considered as a fatal objection.

. . . We conclude that the speed of gravity may provide the new insight physics has been awaiting to lead the way to unification of the fundamental forces. . . . Moreover, the modest switch from SR to LR [Lorentzian Relativity] may correct the "wrong turn" physics must have made to get into the dilemma presented by quantum mechanics, that there appears to be no "deep reality" to the world around us. Quantum phenomena that violate the locality criterion may now be welcomed into conventional physics."[6]

Gravity appears to propagate at extreme super-luminal velocities[6]. It may safely be concluded that the logical inconsistencies of Bell's theorem and inequalities do not in fact preclude non-locality in its super-luminal aspects. Mayants also comes to the conclusion that photons can vary from c, and names fundamental sub c non-zero rest mass expressions of EM: emons.

Next recall the famous wave function collapse of the double-slit experiment. A photonic interference pattern 'collapses' if measured to become something more closely akin to a single particle. This is traditionally ascribed to the effect of "measurement/observation." What can be made of this paradoxical anomaly where the observer affects the observed to induce wave function collapse and perhaps even 'create reality' ?

Measurement or observation are not the bottom of the process; they are but second order descriptions. Wheeler was highly insightful to posit information at the very deepest level of physical reality. Observation and measurement in terms of a primary informational dynamism then represent: *Informational Exchange.* Information affects physical form. Indeed this is true also in biology, not surprisingly, as biology and its relation to chemistry are founded on a primary physical basis with information at the deepest level[7]. The paradox appears as such, only because the primary role of information and its exchange, which clearly affect form/outcome, has not been understood. Now, it may be seen that there is no wave function collapse in the usual sense; the interference pattern is a complete outcome formed of photons, and the 'collapsed' expression is again a different complete outcome formed of photons, both being not in any way uncertain or indeterminate, the differentiation between them being a product of informational exchange which is the dynamic at the bottom of both observation and measurement.

To place this in a human perspective, and suggest a few alterations to the Copenhagen interpretation and some of the more radical theoretic anomalies which have gained predominant sway, such as the deeply troubling many worlds hypothesis, or the equally vexing solipsistic implications of observation, ideas so strange as to have one wonder if an electron is there or perhaps the moon if we are not looking, and place all this into proper relation to probability, attention might turn to some of the more puzzling experiments which are now mounting up and deserve to be addressed.

In these experiments, double-slit interference patterns are seen to change due to thought, and random number and event generators which are properly shielded become more organized in their output. These effects are created at close range, and at *very great distances*[8-13]. Is this inexplicable paranormal activity, or perhaps the cognitive result of resolved uncertainty affecting photonic wave expression? Theory allows an answer: No! This is simply the physics of informational exchange.

In the case of gravitation and also of thought as it affects reality it appears that some nonlocal aspect is needed to explain the effects we observe. Recall the unlikely assertion by Popper, that appears quite clearly to confuse abstract and concrete elements, which states that probability fields must be attributed reality. He had observed experimental effects which required explanation, a real physical explanation was demanded to account for observed phenomenon, hence his supposition. It might reasonably be posited that probability is not at the root of physical form but that information is. Hence it might be hypothesized that the field in question is not a probability field, but a non-probabilistic informational field: *the 'bit field.'*

Imagine a simple example of probability: one reaches one's hand into a concealed container to extract a ball or game chip with some particular marking common to a sub-set of the total objects in the container. Probability is used to guess at result prediction, but in fact the hand does not extract the chip or ball by way of probability, each concrete case is that of selecting one particular concrete object, probability is invoked only to allow prediction under uncertain conditions of human observational constraint, and hence reflects a limit in our available knowledge, not the basic dynamic of the system which is not probabilistic but specific. It has been clearly understood and articulated in previous articles[14] that the nature of human perception is by phenomenological necessity and anatomical analysis understood to be entirely probabilistic. Probability is a valuable and necessary consequence of our human limits. It is a second order method and not a descriptor of underlying processes, but instead an admission of our human limits in defining those processes. Wave function is a necessary abstraction.

With this in mind, the many seemingly paradoxical aspects of quantum theory under the current interpretation may now be reassessed:

There are no many worlds, as the wave function is a probability distribution, an abstract thing which does not require its unrealized aspects be accounted for in some imaginary other world, for all outcomes are complete in and of themselves. There is no uncertainty or wave/particle duality endemic to physical dynamism, those are aspects not of the system at its lowest level of operation, but reflect our human limits which are revealed in attempting to ascertain the same. Uncertainty is the product and province of human cognition and phenomenology, not external reality. Human mental effects upon physical reality including observation/informational-exchange entirely within the sphere of mentation are revealed in experiments referenced above to yield a very slight but demonstrable impact on physical systems. It appears that there is an experimentally demonstrable and specific place for human consciousness in quantum theory, but not the solipsistic one supposed. It may rightly be concluded that human observation in no case creates an electron to observe it, any more than human observation itself might create the moon. The appearance of probability alteration in experiments with human mentation indicates specific informational exchange over some actual medium, perhaps one such as the proposed 'bit field.' It may be concluded that

*The wave function itself represents an abstract probability distribution, signifying the **possible effects** of a potential REAL alteration in systemic informational allocation.*

The fact that subatomic particles demonstrate some fuzziness and do not behave as virtual little golf balls but in a way more akin to a wave packet, is then not due to the fact that the particle is somehow wave-like or uncertain, but because it is a process, a *specific* process which is informationally interactive, as are the larger emergent structures which they compose en masse.

Future questions:

1. Is the implied connection between gravitation, informational exchange and refraction testable in quantum experiments? Clearly alteration in refractive index can account for faster than *c* propagation speeds for light. If informational exchange over a 'bit field' accounts for the super-luminal aspects of gravity, and gravitation can be accounted for in its effects upon light by way of alterations in the refractive index as suggested

above[6], then an experiment could be derived where the hypothesis is tested. Hypothetically: *Micro-gravitational effects created through interactive informational exchange alter refractive conditions yielding specific patterned allocations within the experiment thereby determining the outcome.* Can these theoretical postulations be tested?

2. Is probability at the basis of physical reality or is non-probabilistic information? Does the uncertainty relation signify an endemic systemic aspect, or a human phenomenological limit in epistemology?

3. Is there a realistic interpretation of quantum theory which allows for the unification of gravitation in an informational model based around the empirical necessity of other than *"c"* electromagnetic propagation speeds and experimentally observed nonlocal aspects? Can a quantum model be derived without uncertainty or duality by way of accepting a central tenet of 'informational gravitation'?

4. Is paradox endemic to reality, or is it simply a misunderstanding based on improper assumptions which confuse abstract and real elements?

5. Is it possible to create sound physics based on a constant vacuum propagation value for c? Does Lorentzian relativity offer an alternative?

6. Is the 'bit field' real?

7. Can clear and evident effects of informationally encoded photons on morpho-functional outcomes in biological systems[7] be taken as a correct model for a system-wide common informational basis in physics?

8. Does information theory offer us the elusive prize and connect together gravitational effects with quantum theory by placing informational gravitation as a quantum basis?

9. It appears that the 'bit field' (previously aka the temporal field) *mediates specific entangled relational properties and strength* such as that between a mind and an object or *between gravitationally interactive bodies*, and recent experiments and theories have concluded entangled evolution to be the source of time. Then, could the 'bit field' provide a specific mechanism for temporal/gravitational effects such as gravitational time dilation and others?

Concluding comments.

Our understanding of what light is and is not depends crucially on two interrelated things – experiments carried out meticulously and the theory used to interpret those experiments. As discussed earlier, in his work, when Mayants wishes to talk of particles, he carefully distinguishes between what he calls the *concrete* particles of experiment and the *abstract* particles of theory. This is a rather clever and useful distinction to consider. Experiments are involved with actual reality; theory is always the product of the human mind and, as such, only ever attempts to picture reality rather than be reality. As a result of incorporating this distinction into his reasoning, he has reignited the debate concerning the nature of light – is it merely waves or is a beam of light composed of a huge number of particles? His theoretical calculations look again at the uncomfortable notion of wave/particle duality and show that a particulate theory is capable of describing all events concerning light previously thought to be purely wave phenomena. This proves to be particularly interesting given the recent resurrection of the atomistic view of matter[15] in which everything is fundamentally composed of indivisible particles and void. Interesting because that theory also reduces a light beam to a stream of particles.

As a general point emerging from this discussion, it is worth realising that Mayants also seems to be indicating that great care must be taken when considering any so-called thought experiment. Any theoretician contemplating a physical problem essentially builds a model in his mind to describe the system involved. He then uses well-established techniques, often involving mathematics, to try to understand and explain the original phenomenon. In any thought experiment, the entire process of conceiving an experiment and carrying it out is confined to the mind of the person concerned. There is not any direct contact with physical reality such as is experienced by the experimenter in his laboratory. It seems that Mayants distinction between concrete and abstract particles as discussed here may have farther reaching consequences for future scientists.

Whatever the public view of many might be, these considerations primarily due to Mayants, together with some factors already well-known but highlighted publicly by Santilli, must reawaken the wave/particle duality debate because, in truth, they cast real doubts on that interpretation having much, if any, present day validity.

References.

1. L. Mayants, 1984, *The Enigma of Probability and Physics*, Reidel, Dordrecht.

2. L. Mayants, 1989, Annales de la Fondation Louis de Broglie, **14**, 177-189.

3. L. Mayants, 1994, *Beyond the Quantum Paradox*, Taylor & Francis, London.

4. R. M. Santilli, 2006, *Isodual Theory of Antimatter*, Springer, Dordrecht.

5. Quantum Mechanics without "The Observer" Karl R. Popper in: *Quantum Theory and Reality,* Volume 2 of the series *Studies in the Foundations Methodology and Philosophy of Science* pp. 7-44 Springer-Verlag, Berlin.

http://citeseerx.ist.psu.edu/viewdoc/download?doi=10.1.1.473.23&rep=rep1&type=pdf

6. Van Flandern, T., 1998, The speed of gravity what the experiments say, *Physics Letters* A **250**, 1-11

http://www.sciencedirect.com/science/article/pii/S0375960198006501

7. Norman, R., Dunning-Davies, J., Heredia-Rojas, J. A., Foletti, A., 2016, Quantum Information Medicine: Bit as It— the future direction of medical science: antimicrobial and other potential nontoxic treatments. *World Journal of Neuroscience*, 6. 193-207. http://dx.doi.org/10.4236/wjns.2016.63024

8. Jahn RG, Dunne BJ, Nelson RG, Dobyns YH, and Bradish GJ., 1997, Correlations of random binary sequences with prestated operator intention: A review of a12-year program, *Journal of Scientific Exploration*, **11**, 345-367.

9. Radin D, and Nelson RD., 1989, Evidence for consciousness related anomalies in random physical systems. *Foundations of Physics*, **19**,1499-1514.

10. Radin D, Taft R, and Yount G., 2010, Effects of Healing Intention on Cultured Cells and Truly Random Events. *The Journal of Alternative and Complementary Medicine.* **10**, No 1, pp. 103–112

11. Radin D, Michel L, Galdamez K, Wendland P, Rickenbach R., and Delorme A., 2012, Consciousness and the double-slit interference pattern: Six experiments, *Physics Essays*, **25**, 157-171. doi: 10.4006/0836-1398-25.2.157

12. Radin D, Michel L, Johnston J, and Delorme A., 2013, Psychophysical interactions with a double-slit interference pattern. *Physics essays*, **26**, 4, 553-556 http://dx.doi.org/10.4006/0836-1398-26.4.553

13. Tressoldi P, et al., 2014, Mind-matter Interaction at a Distance of 190 km: Effects on a Random Event Generator Using a Cutoff Method. *NeuroQuantology,* **12**, No. 3, 337-343.

14. Norman, R., 2015, Quantum Unconscious Pre-Space: A Psychoanalytic Neuroscientific Analysis of the Cognitive Science of Elio Conte—The Hard Problem of Consciousness, New Approaches and Directions. *Neuroquantology*, **13**, 487-501. (http://dx.doi.org/10.14704/nq.2015.13.4.869)

15. E. G. Haug, 2014, *Unified Revolution*, EGH Publishing, Norway.

8. Some Possible Links Between Drugs and Violence.

Richard Lawrence Norman.

Abstract:

'Conventional wisdom' within the field of medical psychiatry as evidenced by a great many practitioners, is that mental imbalance is most effectively addressed with drugs. New demonstrably efficacious compounds are supported with studies and touted as a primary therapeutic interventional pathway for the treatment of illness. After study and direct observation, I have deduced several specific facts and relations which are not acknowledged within the current field of psychiatry and may constitute a surprising and consistent factor in the rash of unexplained social violence and rampage killings which have become so prevalent. The specific theory, sociopathic patterns, pathogenic aetiology, neuroscience and psychology are revealed which underlie this new rash of social pathology.

Theoretic introduction. 5-HT and repression. The key Indoleamine—our unconscious gateway; of civilization, creativity and hell.

Today, we are in a unique position. For the first time in pharmacological history we have achieved a level of specificity which has hitherto been inaccessible, and many hands are to be shaken and bows taken. SSRI drugs have specifically targeted the re-uptake of a single neurotransmitter, 5-HT (5-Hydroxytryptamine), and made a new level of neuro-chemical specificity, and individual targeted therapeutic activity available to millions. So, let us assess this new discovery, which I can attest by my personal experience, is most efficacious. As a sufferer of debilitating OCD for many years, you can rest assured in the knowledge that these drugs do work, and are effective in preventing the symptoms of OCD. Those who claim that these potent drugs are ineffective, and have no use or benefit, *are lying to you*. The drugs work. A skilled clinician, should you be lucky enough to find one, can prescribe them in the correct dosages to control your symptoms. Those will be high doses. Now that that is settled, you should also know another fact: Those studies [examine who funds studies] and sources, which claim these drugs are easily withdrawn, and the resultant symptoms are fairly short lived, most definitely and assuredly, *are lying to you*. Please know the fact: SSRI drugs administered in the proper high doses for disorders such as OCD over long periods, cause permanent damage to the repressive system—

Repression is 5-HT dependent (Norman, 2009, 2010, 2011, 2013, 2013*a*).

That dry statement, "Repression is 5-HT dependent," has consequences and specific implications, some unexpected, which have changed in ways both positive and otherwise, the entire landscape of psychology. Now, old and vital questions have been answered, and the question of the existence of unconscious fantasy (Talvitie & Ihanus, 2005) and its influence on behavior and the transference have finally been lain to rest (Norman, 2011, 2013, 2016, 2016a). First, I will begin with a general assessment of the specific ontogenetic manifestations and neuroscientific mechanisms involved.

The various transformations of illness which parallel the reduction in repressive functioning as SSRI withdrawal occurs, are necessary symptomatic products of the return of repressed material to consciousness (Freud, 1896, p.170 [first usage of the phrase]), and demonstrate the common defensive and purposive mechanisms of neurotic and psychotic illness (Freud, 1896; Norman, 2010, 2011, 2013). The result is surprising, not because it supports the Freudian idea of all such illness being manifestations of defence rather than random imbalance, but, because the usual barriers which favor one illness over another, the "predispositional" factor itself, seems to have been cast aside (Norman, 2013). This is easily accounted for if we remember that this is an artificial neurosis/psychosis, not a typical one, and hence, must be assessed on its own footing. The mechanism by which it and its transformations are created, is clear: a relative reduction in 5-HT in the synaptic cleft due to the resumption of normal 5-HT re-uptake, and a resultant wholescale reduction in repressive function (alongside concurrent effects due to any physiological damage from extended treatment). With repressive function permanently impaired, what were predispositional influences favoring illnesses which are dependent upon high levels of repressive functioning such as OCD, are now exposed in their internal construction, repression peeled back, and the core of hysterical illness laid bare. The resultant hallucinatory hysterical psychosis, demonstrates little symbolic distortion of its reactive components, which may be assessed quite directly.

This psychosis, which can be reverse engineered to allow us access to undistorted unconscious content in some cases, has specific concurrent manifestations regarding perception. Repression and the unconscious have subsumed under their functioning, not only a temporally "passive" role (retroactively defining reality) *in relation to the level of perceived conscious input* of previous externally derived experience, functioning not only in the familiar role as a receptacle for containment, affective dampening, dynamic removal and allocation to experience of *preexisting* internal (interoceptive) unconscious stimuli such as memories and fabricated conglomerations such as unconscious fantasies via transference, but also an active one as well. This active real-time repressive function whereby all of perceptual experience has its energetic incoming presentation reduced, *actively repressed* in large measure into the unconscious *as it happens*, I have called: The Active Unconscious (Norman, 2010). Although the concept was conceived before I read the Freud, this is a more functionally connected and useful extension of Freud's stimulus barrier (Freud, 1920, p. 27). This reduction in the ability to partly repress the full force of *external* experience (exteroceptive increase), which runs in close tandem with the concurrent loss of ability to repress the influence of our *internal* perceptions stemming from the unconscious (interoceptive increase), form the full measure of repression proper, and are inexorably joined, rising and falling together in their level of functioning in direct and dependent relation to the increase or decrease in systemic levels of 5-HT.

There is ample neuroscientific evidence to support and explain this mechanism, by virtue of which I myself have been transformed from an extrovert who wanted only more and more intense stimulus, performing before larger and larger crowds, into an introvert, a man who is overwhelmed by natural beauty, weeps openly and often, and feels a sunbeam on his flesh with the same shuddering amazement I used to gain only by way of the most extreme and daring behavior. It is as if the very most basic and fundamental of psychical relations has been altered, and not in any subtle way! The idea that SSRI drugs are specific in their action, is both laughable, and utterly mistaken. These drugs target one of the most evolutionarily ancient systems in the brain, as is evidenced by the central location of the serotonin producing nuclei, which dispense 5-HT to no less than 15 receptor types (Panksepp, 1998, p.111). The list of behavioral functions which *do not* involve brain serotonin is quite short, and can be represented by a single digit: Zero. Yes, 5-HT is so basic, its functions so diverse, we can say: 5-HT is involved...*in everything* (Panksepp, 1998, p. 103). The psychical effects of serotonin depletion and supplementation are no mystery, and neither are its general systemic effects:

Jaak Panksepp, founder of the burgeoning discipline known as Affective Neuroscience, has made one of the most profound, direct and reliable contributions to our knowledge of human and animal neural affective dynamics, from both evolutionary and biological perspectives. This careful and detailed researcher, has by way of experiment and observation come to certain conclusions about the role of brain serotonin in brain processes and behavior.

Firstly we read in Panksepp (1998) [citation form altered]:

"There are good reasons to believe that this system mediates a relatively homogeneous central state function. All motivated and active emotional behaviors including feeding, drinking, sex, aggression, play and practically every other activity (except sleep), appears to be reduced as serotonergic activity increases (Coccaro & Murphy, 1990; Jacobs & Gelperin, 1981) (Panksepp, 1998, p. 111)."

The fact that 5-HT has *some* receptors which increase anxiety, is in my view, not at all inconsistent with the role of 5-HT mediating repression, as anxiety is in many cases the causal instrument by which repression is instated (Freud, 1926; Brenner in Rickman, 1957; Norman, 2010, 2011). We read a general description of the effects of brain serotonin on mental stimulation of both interoceptive and exteroceptive origin, which makes some good sense of the relation between 5-HT and repressive function both "passive" and "active" as previously described. Description from a diagram of 5-HT pathways (Panksepp, 1998):

"Serotonin. Function: reduces impact of incoming information and cross talk between sensory channels" (p. 107). As to the resultant behavioral modifications when brain 5-HT is reduced, (which closely parallel those of REM deprivation): ". . . such animals are behaviorally *disinhibited:* they are more active, more aggressive, hypersexual, and generally exhibit more motivational/emotional energy. . . In short, they appear to be manic." (p. 141).

And lastly, we read:

> "In general, it seems that one higher cerebral function of brain serotonin is to sustain stability in perceptual and higher cognitive channels. When this constraint is loosened by a global reduction of 5-HT activity, the probability of information from one channel crossing into another channel is increased. Thus a mild reduction in brain serotonin activity may be an important ingredient for the generation of new insights and ideas in the brain, while the sustained reduction of serotonin might lead to chaotic feelings and perceptions, contributing to feelings of discoherence and mania.
>
> In sum, perhaps it is this loosening of sensory-perceptual barriers between different brain systems that characterizes dreams, hallucinations and the florid phases of schizophrenia, as well as normal creativity. . . it is worth noting that just as low brain serotonin characterizes the dream

state, it also promotes heightened emotionality, both positive and negative. It is a neurochemical state that leads to impulsive behavior in humans (Halperin et al., 1994; Linnoila, et al., 1983; Roy et al., 1988), even ones as extreme as suicide (Asberg, et al., 1976; Brown et al., 1982; Coccaro, 1989). Probably the most striking and replicable neurochemical finding in the whole psychiatric literature is that individuals who have killed themselves typically have abnormally low brain serotonin activity." [Panksepp, 1998, p. 142]

I hope the exact and full implications of this statement are becoming more clear: "Repression is 5-HT dependent." In less technical language you can imagine brain 5-HT, its particular manifestations and effects to be better summed in this less precise but more descriptive phrase: 5-HT is the lid on hell. So now that modern pharmacology has removed the blinders, and allowed us direct access into the forbidden ugliness which is within all mankind, this hidden fuel of his ascension and decline, for all of sublimation and depravity are found within this secret—*let us look.* We will see the main of Freudian theory, this hideous and unflattering picture of inner reality... is essentially correct. However, the situation does not unfold quite as the effects do with animals, and indeed, an SSRI withdrawal subject would wish for a blessed mania to quell their pain, for unlike animals, we have super-ego, and super-ego is masochistic, as a punitive garrison set up within personality (Freud, 1930, pp. 123-124; Norman, 2013a). When we add a punitive super-ego wish to an id wish with reduced repression we have the exact description of the dynamic which creates hysteria proper (Freud, 1915, pp. 180-185). I hope it is now becoming clear to the reader, why, SSRI withdrawal encourages *hysterical hallucinatory psychosis.*

Drugs and Murder – a possible link:

We are in an age which is fraught with change, some positive and some less so. It seems as if the basic fabric of our culture has torn, as if a qualitatively new and distinct rash of horror and criminal activity has overtaken this age and defined it: the rampage killing, a new sort of crime which appears to defy explanation, but do be sure this is false, and an explanation is at hand. Indeed, these crimes are nothing if not utterly predictable. I will offer up my theory as to the psychological mechanism involved here.

So what has changed? Why are there so many rampage killings, now as never before appearing with such alarming frequency, school shootings, mall murders, movie theatre massacres and the like? There have always been guns in our American society, always so very many guns, but no, these shootings and murders are appearing on a scale never before seen. Ergo: the mechanism must lie elsewhere. There have been neglectful parents and bad children throughout history, so very many bad parents and ugly mean spirited children, but no, these crimes are so tragic and only now, so prolific, so violent and today so much more frequent. Ergo: the mechanism must lie elsewhere. The answer is, although belatedly, becoming clear. I will list but a few cases with partial pharmacological histories and then analyze the connecting factor:

John Shick, 2012, age 30, killed one injured six, was shot by police. Nine different anti-depressants were found in his apartment.

Hammad Memon, 2010, age 14, Shot and killed a student at school. He was taking the SSRI Zoloft.

Christopher Wood, 2009, age 34, cut and shot his wife and three children and committed suicide. He was taking the SSRI Paxil.

Jason Montes, 2009, age 33, killed his wife and shot himself. He was taking the SSRI Prozac.

Steven Kazmierczak, 2008, age 27, killed five, wounded twenty-one then killed himself. He was taking the SSRI Prozac.

Jeff Weise, 2005, age 16, killed his grandfather, grandfather's girlfriend, then drove to the high school, killing seven, wounding five and shooting himself. He was taking the SSRI Prozac.

Doug Williams, 2003, age 48, shot fourteen co-workers, killing six before turning the gun on himself. He was taking the SSRI Zoloft.

Eric Harris, 1999, age 18, along with Dylan Klebold, age 17, shot and killed twelve students and a teacher, wounding twenty-six others before killing themselves. Harris was taking the SSRI Luvox; Klebold's medical records are unavailable.

Kip Kinkel, 1998, age 15, shot his parents to death with a rifle, went to school and open-fired in the cafeteria, killing two and wounding twenty-five. He had been taking the SSRI Prozac.

So let me state at the outset that nothing could be more puerile, reactionary and short-sighted than to condemn an entire class of worthy drugs which are potentially so beneficial, like SSRI drugs, of which Prozac is the most prominent representative. When properly prescribed these drugs do vital and good work. However, these drugs work in specific ways which entail risks. These risks are utterly predictable and have largely been ignored. Do note the similarity in behavior connecting the above mentioned crimes which all entail a violent outburst and then, in many cases end in death by police or suicide. This pattern is created as a psychological function of the neuro-chemical effects of SSRI therapy, tolerance and withdrawal, as these factors interact in specific and predictable ways. Although websites such as SSRIstories.com and the Citizens Commission on Human Rights website at cchrint.org offer information correlating these crimes with SSRI use and withdrawal, there is not enough information specifying the psychological mechanisms which yield these behavioral effects. I will offer a general analysis of those mechanisms here.

Conscious vs. Unconscious: To understand these factors, we must first understand the basics of unconscious psychology. When an external threat is perceived, we run away or fight. However, the situation is different if the threatening factor comes from within us. Our own ideas, memories and thoughts can be every bit as dangerous to us, and to society, as an external enemy. As we grow up, we learn to control our aggressive and sexual instincts. These ideas and instincts are never truly gone, and can be seen to "reappear" in certain circumstances, such as under conditions of painful deprivation, madness and war, where every murderous human instinct can be seen to reemerge. These instincts then, have never disappeared, rather, they have been repressed, and made unconscious. Society is built upon the bedrock of repression and the unconscious. Psychology informs us, that as these internal instinctual threats return to consciousness, we become ill. In the language of Freudian psychology: symptom formation is a function of the return of the repressed.

I have discovered that SSRI drugs positively affect mental processes by reinforcing repression: repression is 5-HT (5-Hydroxytryptamine) dependent, and SSRI drugs increase 5-HT in the neuronal network by preventing re-uptake of the neurotransmitter in the synaptic system. [I will refer you to the latest edition of Goodman and Gilman's *The Pharmacological Basis of Therapeutics* for a complete description of the neurochemistry involved in the effects of SSRI therapy.] By increasing the amount of 5-HT in the neural system, and preventing the repressed from entering consciousness, they quell mental illness. However, as is the usual case with drugs, tolerance develops and functions as partial withdrawal, and, many patients do, in fact, withdraw from these drugs. In this instance, the effect is reversed, and repression is circumvented, allowing unconscious material to enter consciousness. So the drug that helps by way of reinforcing repression, causes illness as repression is reduced by way of tolerance or withdrawal.

This reduction in overall repressive function manifests itself as an unusual artificial hysterical psychosis, where both aspects of repression are circumvented, amnesia, and distortion via compromise-formation symbolism. If the dose is high, and the term of treatment long, upon withdrawal the effect is severe. In delusion, the psychotic is afforded a level of protection, as his delusion is a sort of distortion, a symbolic transformation of the wishes and/or mnemic experiences which are returning to consciousness and creating his illness (Freud, 1911, pp. 1-82; 1924, p. 151). Now, in SSRI withdrawal, even this most basic protective function of dream and delusion is defeated, and the most energetic and severe of unconscious material can gain direct and unfettered access to consciousness, free from any distortion. The effect to the ego is absolute and certain: damage of the most severe sort. Super-ego/ego is directly exposed to the most toxic unconscious contents, and its repressions further disintegrate, further revealing the very most energetic and highly disturbing hidden ideations. Sleep, in some cases, may be curtailed to as little as three hours or less a night. Soon, hallucination completes the picture, and a new sort of even more dangerous and severe psychosis is seen to emerge.

I will briefly traverse a secondary avenue of interest before completing the picture. Although the technical, psychological and medical information associated with these drugs is substantial, the fact that repression itself is affected to create behavioral effects has been utterly ignored. *The fact that repression is 5-HT dependent has not been articulated.* The result is clear: as repression is decreased through SSRI withdrawal, two things can be counted upon:

1. A mental illness, whatever its relation to repression, be it defined by the deepest repressions such as OCD or not, *will* be converted into an hysterical illness as hysteria is formed through the return of repressed unconscious contents under *low levels of repression* (or I postulate *perhaps* trigger the emergence of schizophrenia if the subject is predisposed). That is why hysterics demonstrate conversion hysteria, a bodily innervation of opposing wishes, in lieu of more typical repressive means (Freud, 1915, pp. 184-185), or anxiety hysteria, a common hysterical reaction in children, who have yet to develop a high level of repressive function (Freud, 1909, pp. 1-149; 1915, pp. 182-184).

2. As hysterical illness is formed through SSRI withdrawal, the job of analysis is made much easier, as unconscious ideations which are pathogenic are more easily accessed (Norman, 2011). It should be noted that these contents are likely to reveal themselves as negative transference, which although

shunned in modern analysis, is in fact the key to un-riddling the puzzle.

Now we must add but one more bit of information and the analysis will be clear. Our aggressive drives are deeply repressed. These drives are repressed as a function of conscience, of guilt and super-ego, which acts as a conscious "reaction formation," an opposite which fills up consciousness as a replacement, a substitute for the repressed drive (Freud, 1923, p. 56). Sadism, violence used to control an object with no concern for that person or object, is chief among those drives we repress. The unconscious is filled with sadism. When we add guilt to a sadistic stream of great force and potency, the sadism "turns round" on the subject and becomes masochism, the chief representative of the death instinct (Freud, 1919, pp. 193-194; Norman, 2011, p.116).

Guilt + Sadism = Masochism. Now the analysis is plain:

A mentally ill person is placed on SSRI drugs that function to enforce their repressive facility which is failing and creating illness as their overly potent repressed drives return to consciousness. Soon the drug fails to maintain its effect as tolerance ensues, or, the subject withdraws from the drug. Now, repression is defeated, and unconscious content becomes conscious in its most toxic, direct and uncensored form. The subject identifies with his sadistic thoughts which present with such energetic force, as to be utterly irresistible. Once his hatred is spent, the guilt he feels for his actions is revealed, and added to his freed conscious sadistic drives to form masochism, and suicide, often suicide by way of police intervention. The psychology is utterly obvious, and, predictable. (Of course, the more likely result is suicide alone, and the above mentioned pattern of behavior is formed in those cases where sadistic ideation has obtained an energetically predominant place in the mental architecture).

Now imagine the combat veteran, trained in the art of killing, he returns to our shores, a hero, but ill for his service, ill for the guilt of killing. He is prescribed an SSRI drug, and feels better. Soon he tires of the debilitating side effects, and discontinues therapy. Can you see it? What will become of him then? What will become of us? If you are taking one of these drugs, I urge you not to stop. If you do stop, do it slowly, so very slowly, and be careful. These people who kill are not so different than any of us, in fact, any of us could be one of them. Although perhaps differing in intensity and proportion, all of us have these drives... every single one. The only difference is that we can contain them, and can not see them, can not see this part of ourselves. Perhaps the only real difference between one of these killers and one of us, is a misfortune of human honesty, in that they, are unfortunate enough to know a little too much— of themselves. So when you wonder what separates a mad killer from one of us, you may be surprised to learn the difference may be as small as a single question - a question, of human honesty.

Concluding remarks:

The pharmaceutical industries and their lobbies spent ~$235,107,261 in 2015 in support of their interests. These financial giants pour many millions of dollars into advertising and costly informational distribution aimed directly at patients and physicians. However, the number of well funded studies examining the long term effects of SSRI treatment are scant. Not surprisingly, the result of studies which have been conducted are in keeping with my personal findings and research, and indicate permanent damage associated with SSRI use in depression (El-Mallakh et al. 2011). Other trustworthy researchers and doctors have found the same (Breggin, 2011). To my knowledge no studies are available detailing the long term damage associated with the very much higher doses used to treat populations with OCD. Two of the most widely used drug types in the treatment of mental illness, antipsychotics (including the newer 'atypical antipsychotics') and SSRI drugs, have both been scientifically demonstrated to cause permanent damage: tardive dyskinesia and tardive dysphoria respectively. The consequences extend past the personal lives of those affected, and influence society at large. Therefore I wish to suggest these possibilities:

1. Studies which are not funded by the pharmaceutical industries must be conducted which spell out the frequency and level of damage incurred through *all the SSRI dosage levels currently advised in treatment regimens for all conditions and populations treated.*

2. That information should be *actively distributed* to patients and doctors and included in product advertising and labeling where it is made plain in large typeface.

3. Serious consideration must be paid to new approaches which allocate potentially damaging drugs a safer place as a third tier treatment option, and serious consideration and priority given to other more healthful modes of treatment such as talk therapies and others, which may then replace potentially harmful drugs as primary first tier interventionary tools in the treatment of mental disease.

References:

Asperg, M., Traksman, L., & Thoren, P. (1976).

5-HIAA in the cerebrospinal fluid:

A biochemical suicide predictor?

Arch. Gen. Psychiat. 33: 1193-1197.

Breggin, P. (2011) New Research: Antidepressants Can Cause Long-

Term Depression. Retrieved from: *Huffington Post* on line:

http://www.huffingtonpost.com/dr-peter-   breggin/antidepressants-long-term-depression_b_1077185.html

Brown, G. L., Ebert, M. H., Goyer, P. F., Jimerson, D. C.,

Klein, W. J., Bunney, W. E., & Goodwin, F. K. (1982).

Aggression, suicide, and serotonin: Relationship to

CSF amine metabolites. *Am. J. Psychiat.* 139: 741-746.

Coccaro, E. F. (1989).
Central serotonin and impulsive
aggression. *Br. J. Psychiatr.* 155: 52-62.

Coccaro, E. F., & Murphy, D. L. (eds.) (1990).
Serotonin in major psychiatric disorders. Washington,
D.C.: American Psychiatric Press.

El-Mallakh, R. S., Gao, Y., Jeannie Roberts, R. (2011)
Tardive dysphoria: The role of long term antidepressant use in-
inducing chronic depression. *Med Hypotheses* (6):769-73 doi:
10.1016/j.mehy.2011.01.020
http://www.ncbi.nlm.nih.gov/pubmed/21459521

Freud, S. (1893-1899). *The standard edition of the complete*

psychological works of Sigmund Freud volume three:

Early psychoanalytic publications. London: Hogarth Press.

Freud, S. (1909). *The standard edition of the complete psychological*
works of Sigmund Freud volume ten: Two case histories: 'Little Hans'
and 'Rat Man'. London: Hogarth Press.

Freud, S. (1911-1913). *The standard edition of the complete*
psychological works of Sigmund Freud volume twelve: Case history
of Schreber, Papers on technique, and other works. London:
Hogarth Press.

Freud, S. (1914-1916). *The standard edition of the complete*

psychological works of Sigmund Freud volume fourteen:

On the history of the psycho-analytic movement,

Papers on metapsychology, and other works.

London: Hogarth Press.

Freud, S. (1917-1919). *The standard edition of the complete psychological works of Sigmund Freud volume seventeen: An infantile neurosis, and other works.* London: Hogarth Press.

Freud, S. (1920-1922). *The standard edition of the complete psychological works of Sigmund Freud volume eighteen: Beyond the pleasure principle, Group psychology and other works.* London: Hogarth Press.

Freud, S. (1923-1925). *The standard edition of the complete psychological works of Sigmund Freud volume nineteen: The ego and the id, and other works.* London: Hogarth Press.

Freud, S. (1925-1926). *The standard edition of the complete psychological works of Sigmund Freud volume twenty: An autobiographical study, Inhibitions symptoms and anxiety, Lay analysis, and other works.* London: Hogarth Press.

Freud, S. (1927-1931). *The standard edition of the complete psychological works of Sigmund Freud volume twenty-one: The future of an illusion, Civilization and its discontents, and other works.* London: Hogarth Press.

Halperin, J. M., Sharma, V., Siever, L. J., Schwartz, S. T., Matier, K., Worknell, G., & Newcorn, J. H. (1994). Serotonergic function in aggressive and nonaggressive boys with attention deficit hyperactivity disorder. *Am. J. Psychiat.* 151: 243-248.

Jacobs, B. L., & Gelperin, A. (eds.) (1981).
Serotonin neurotransmission and behavior.
Cambridge, Mass.: MIT Press.

Linnoila, M., Virkkunen, M., Scheinin, M., Nuutilia, A.,
Rimon, R., & Goodwin, F. K. (1983). Low cerebralspinal fluid
5-HIAA concentration differentiates
impulsive from nonimpulsive violent behavior.
Life Sci. 33: 2609-2614.

Norman, R. (2009). *This new day—Self creation: The wisdom of an
idiot.*   O'Brien, OR.: Standing Dead Publications.

Norman, R. (2010). *Mind map: Psychological topography
and an approach to a new creative psychology,
or, the secret of happiness.* O'Brien, OR.:
Standing Dead Publications.

Norman, R. (2011). *The tangible self.* O'Brien, OR.:
Standing Dead Publications.

Norman, R. (2013). Nine Short essays and *Native Psychoanalysis—
a Non-Elliptical Technique*: Necessary Background
Information Basic to Native Psychoanalysis.
The Black Watch: The Journal of Unconscious Psychology
and Self-Psychoanalysis. Retrieved from:
www.thejournalofunconsciouspsychology.com

Norman, R. (2013*a*). Who Fired Prometheus?
The historical genesis and ontology of super-ego and
the castration complex: The destructuralization
and repair of modern personality—An essay in five parts.

The Journal of Unconscious Psychology and Self-Psychoanalysis.
Retrieved from: www.thejournalofunconsciouspsychology.com

Norman, R. L. (2016) The Quantitative Unconscious: A
Psychoanalytic Perturbation-Theoretic Approach to the Complexity
of Neuronal Systems in the Neuroses, *Neuroquantology*, Vol. 14
issue 2 10.14704/nq.2016.14.2.949 **356-368**

Norman, R. L. (2016a) Homeostatic Conductance and
Parasympathetic Basis Alteration: Two Alternative Approaches to
Deep Brain Stimulation in Parkinson's, Obsessive Compulsive
Disorder and Depression. *World Journal of Neuroscience*,
6, 52-61. http://dx.doi.org/10.4236/wjns.2016.61007

Panksepp, J. (1998). *Affective Neuroscience: The Foundations of
Human and* Animal Emotions. New York, NY.: Oxford Press.

Rickman J, (Ed.) (1957). *A general selection from the works
of Sigmund Freud.* New York, NY.: Doubleday.

Roy, A., Adinoff, B., & Linnoila, M. (1988). Acting out hostility
in normal volunteers: Negative correlation with levels
of 5-HIAA in cerebrospinal fluid. *Psychiat. Res.* 24: 187-194.

Talvitie, V., & Ihanus, J. (2005). Biting the bullet:
The nature of unconscious fantasy.
Theory and Psychology. 15(5): 659–678.
DOI: 10.1177/0959354305057268

9. Thoughts Occasioned by Two Announcements

Not too long ago, NASA announced that its Voyager 1 spacecraft had entered a new region between our solar system and interstellar space. All the details may be viewed at

http://www.jpl.nasa.gov/news/news.cfm?release=20 11-372

In this announcement, one of the more interesting comments is that "Voyager has detected a 100-fold increase in the intensity of high-energy electrons from elsewhere in the galaxy diffusing into our solar system from outside". This comment is of interest because, apart from the word 'diffusing', it describes what the electrical model of our universe expects in the virtual cathode region of the solar discharge boundary.

Also, in a separate announcement, it was revealed that a new all-sky map shows the magnetic fields of the Milky Way with the highest precision

http://www.physorg.com/news/2011-12-all-sky-magnetic-fields-milky-highest.html

It was claimed that the origin of galactic magnetic fields remains unknown despite intensive research, although it was seemingly assumed that they are constructed via dynamo processes such as are said to occur – in violation of a well-known theorem due to Cowling incidentally – in the interiors of the Earth and the Sun.

Beyond the Veil Dunning-Davies and Norman

Some years ago, in an entirely different context, Sir Winston Churchill advised people to learn from the lessons of history and, in the present context, it might seem appropriate to follow this advice in astrophysics. Hence, in this spirit, it might be noted that, following the introduction of Newton's mechanical ideas, work still proceeded apace investigating electromagnetic phenomena and this continued at least into the earlier years of the twentieth century, as is evidenced by the contents of J. J. Thomson's book *Electricity and Matter*[1]. However, this book provides but one example to illustrate the very real emphasis on work involving the effects of the electric and magnetic fields, work which, incidentally, constantly sought an explanation for the concept of mass in terms of those forces. However, after those early years of the century, the emphasis seems to have shifted to explanations of phenomena purely in terms of gravitational effects. Considering that it is accepted that much of the matter in the universe is in the form of plasma, this seems a retrograde step. One may only speculate as to why the emphasis of much scientific research changed in this way. However, thanks to people like Birkeland, Alfvén and, more recently, Peratt, work in the areas of electromagnetism and plasma physics has continued.

The work on plasmas and other electromagnetic phenomena has inspired people to examine astronomical phenomena in these terms and this has resulted in the so-called Electric Universe idea as expounded, for example, in the books *The Electric Universe*[2] and *The Electric Sky*[3]. Reading through this material makes one immediately aware that just like accepted theory the electric universe ideas are supported by computer modelling, but it is also able to draw on parallels between astronomical phenomena and plasma phenomena observed in the laboratory. Admittedly, drawing such parallels involves scaling up tremendously but assuming this possible is little different from assuming that laws seemingly applicable here on the Earth are also applicable in the Solar System and, indeed, throughout the universe. At least visually, some of the phenomena observed in the laboratory are very like what is observed by some of the most powerful of telescopes. Electric currents in plasma naturally form filaments due to the so-called 'pinch effect' of the induced magnetic field. Electromagnetic interactions cause these filaments to rotate about one another to form a helical 'Birkeland Current' filament pair and this is very much the structure seen in the Double Helix nebula near the galactic centre; again, the Hubble image of the planetary nebula NGC6751 looks remarkably like the view down the barrel of a plasma focus device. Examples such as these prove nothing but might awaken people to the possibility of alternative explanations for at least some astronomical phenomena.

Much of the laboratory work originated with the work of Kristian Birkeland more than one hundred years ago. It was during his Arctic expeditions at the end of the 19th century that the first magnetic field measurements were made of the Earth's polar regions. His findings also indicated the likelihood that the auroras were produced by charged particles originating in the Sun and guided by the Earth's magnetic field. Birkeland, though, was an experimentalist and is still known for his Terrella experiments carried out in a near vacuum and in which he used a magnetised metallic sphere to represent the Sun or a planet and subjected it to electrical discharges. By this means, he was able to produce scaled down auroral-type displays as well as analogues of other astronomical phenomena. These claims, however, were only vindicated finally by satellite measurements in the 1960's and 70's. To that point in time, his experimental and observational achievements had tended to be overshadowed by the purely theoretical predictions and explanations of the geophysicist, Sydney Chapman. Powerful mathematics seems to have held sway over the more expected techniques of physics – experimentation and observation, with mathematics a mere tool to be used when necessary. This is not to decry Chapman's work but to emphasise the overwhelming importance of the physics when investigating natural phenomena.

Birkeland also showed experimentally that electric currents tend to flow along filaments shaped by current induced magnetic fields. Of course, this confirmed observations of Ampère that indicated that two parallel currents flowing in wires experience a long range attractive magnetic force that brings them closer together. However, as plasma currents come closer together, they are free to rotate about each other. Such action generates a short range repulsive magnetic force which keeps the filaments separated so that they are, in effect, insulated from each other and able to maintain their separate identities. The end effect is for them to appear like a twisted rope and it is this configuration which is termed a 'Birkeland current'. Satellites orbiting above the auroras in the 60's and

70's were able to detect a movement of ions, indicating that electric currents were present. Later missions found quasi-steady electric fields above the auroras following the magnetic field lines, thus lending some credence to Birkeland's claim of the existence of an electric circuit between the earth and the Sun.

However, the so-called Electric Universe is really just an hypothesis, a new way of interpreting known data by using both new and well-established knowledge relating to electricity and plasma. It should be emphasised immediately that, in this new interpretation, gravity still has a role to play but it is a secondary one since the electric force is so much more powerful. A major point to be stressed from the outset is that, in this interpretation of astronomical phenomena, scientists are able to call on evidence from laboratory based experiments to help form and support suggested explanations for a wide variety of phenomena. It has been found that, as explained in more detail in the above-mentioned books, a plasma in a laboratory is a good model for providing possible explanations for many recently observed astronomical phenomena which, in several cases, have puzzled astronomers seeking explanations via more usual routes. This is not to say that gravity is ignored and regarded as irrelevant; rather, the possible effects of the electromagnetic force on astronomical phenomena are investigated while still recognising the importance of gravitational effects. In the electric universe, the gravitational systems of galaxies, stars, moons and planets are felt to have their origins in the proven ability of electricity to generate both structure and rotation in plasma. It is felt further that the force of gravity assumes importance only as the electromagnetic forces approach equilibrium. As has been noted already, great consternation has been caused in astronomical circles by the realisation that gravity, as presently understood, cannot explain much that is observed if the amount of mass available is as now felt to be present. Hence, instead of positing the existence of 'dark matter' or following the path of modifying Newton's well-tried law of gravitation significantly, it is suggested here that the effects of the electromagnetic force be examined to see if, in conjunction with orthodox ideas on gravity, these puzzling observations can be explained. However, returning to the realisation that much of the matter permeating the Universe is in the form of plasma, it might be remembered that these clouds of plasma respond to the well-known laws of Maxwell. Also, as pointed out by Scott in his book[3], another law, formulated by Lorentz, does help explain the galactic speeds alluded to earlier. This law states that

a moving charged particle's momentum (speed or direction) can be changed by application of either an electric field or a magnetic field or both.

This seems a highly likely contributory factor, at least, causing galaxies to rotate as they are perceived to do but would indicate, contrary to the accepted view, that gravity has less to do with things than has been thought. However, it should be noted that nowhere is it being suggested that Newton's law of gravitation is in error; it is simply being suggested that, in deep space where everything swims in a sea of plasma, the Maxwell –

Lorentz electromagnetic forces dominate over those of gravity.

It might be remembered also that the Lorentz force alluded to here changes a charged particle's momentum and that change is directly proportional to the strength of the magnetic field through which the particle is moving. Further, the strength of a magnetic field produced by an electric current is inversely proportional to the distance from the current but the gravitational force between stars is inversely proportional to the *square* of the distance. This well-known difference between the two forces could lie at the heart of the problem of the galactic rotation curves; certainly it seems an avenue worth exploring further, especially considering the fact that more and more space missions are indicating that electromagnetic forces are distributed more widely throughout space and are, of course, many orders of magnitude stronger than gravitational forces.

As well as a great many laboratory experiments being performed to establish plasma properties[4] , it has been shown also, using the Maxwell and Lorentz equations, that streams of charged particles, such as are found in the intergalactic plasma, will evolve into the familiar galactic shapes under the influence of electromagnetic forces. The results fit extremely well with the observed velocity profiles in the galaxies and all this with no recourse to missing mass. Much of this simulation work has been carried out by Anthony Peratt and is reported in various issues of the IEEE Transactions on Plasma Science.

References.

1. J. J. Thomson, *Electricity and Matter,* Westminster: Archibald Constable & Co., 1904.

2. W. Thornhill and D. Talbott, *The Electric* Universe, Portland, Mikamar Publishing, 2002.

3. D. E. Scott, *The Electric* Sky, Portland, Mikamar Publishing, 2006.

4. A. Peratt, *Physics of the Plasma* Universe, New York, Springer-Verlag, 1992.

10. A Discussion of Structure and Memory in Water.

Introduction.

People have speculated for some time over whether substances, such as water, actually have a memory. However, it was in 1988 that a truly staggering article appeared in the journal *Nature* purporting to report the experimental observation of this property assumed by many to be merely an attribute of animals, particularly humans. The article in question[1] by a team, headed by Dr. Jacques Benveniste, claimed to have observed that extremely dilute biological agents were still capable of triggering relevant biological systems. In fact, they even claimed this to be so in the absence of actual physical molecules of the agents concerned. Some of the experiments had been reproduced in laboratories other than Benveniste's and members of these laboratories co-signed the article. However, as has been noted in a popular book on homeopathy[2], this article "provoked a flurry of comment and resulted in the rerun of the experiments under the 'scientific' eyes of a fraud detector, a journalist and a magician." Presumably, by 'a journalist', the writer of this book meant the editor of *Nature*, but the person concerned was by training a physicist and might have been expected to have had some elementary knowledge of information theory and that it had been applied to physical systems. Although a relatively old subject in its own right at that time, information theory had been coming into physics via such books as that of Brillouin[3]. It might have been thought by some that this fact would have introduced a more cautious note into some of the condemnation of Benveniste's work.

The article itself appeared in the issue of the journal for 30th June 1988 and the ensuing furore was such that the then editor of *Nature* summed up his reading of the situation and called a halt to further correspondence in the issue of 27th October 1988, after allowing Dr. Benveniste a chance to answer his critics. What really caused the furore? The answer is best summed up by the 'Editorial Reservation' which appeared with the original article. This said that "readers of this article may share the incredulity of the many referees who have commented on several versions of it during the past several months. The essence of the result is that an aqueous solution of an antibody retains its ability to evoke a biological response even when diluted to such an extent that there is negligible chance of there being a single molecule in any sample. There is no physical basis for such an activity." In the later commentary, attention was drawn to the fact that one of the concerns of the editor of *Nature* was that the publication of the paper was "certain to excite the interest of the homeopathic community". Given this, therefore, it is surprising the article ever appeared in print, but appear it did even though it was stated there was no physical basis to explain the claimed phenomena.

It is this final statement which is now called into question with the appearance of an article purporting to give the biophysical basis of the Benveniste experiments[4] and it is the purpose of this note to draw attention to this work which could be of vital importance in helping establish the scientific validity of homeopathic remedies within the medical fraternity as a whole.

Theoretical background.

The basis of information theory is now well-established. Following the approach of Brillouin[3], if P denotes the number of states in a system, then the information memory capacity (denoted by I) in 'bits' is defined to be

$$I = \ln P,$$

where, if a problem is considered with N different independent selections, each corresponding to a binary choice (0 or 1), the total number of possibilities is

$$P = 2N$$

and so, the information is

$$I = N\ln 2.$$

Alternatively, the entropy function of statistical thermodynamics is given by

$$S = k\ln P,$$

where k is Boltzmann's constant.

It follows that, for the above expression for P,

$$S = k\ln(2^N) = kN\ln 2.$$

Further, it may be noted that the first and second laws of thermodynamics may be combined into the equation

$$dU = TdS + d'W,$$

where dU denotes the internal energy, T the absolute temperature and $d'W$ the work done on or by the system. In terms of memory capacity, this becomes

$$dU = (kT\ln2)dN + d'W$$

and it is seen immediately that the energy required to add one bit of memory to the system is given by

$$kT\ln2 = \frac{\partial U}{\partial N}$$

where the partial derivative is evaluated with the work term held constant.

It might be noted that heat capacity is necessarily a positive quantity[5] and, therefore, this last equation leads to the realisation[4] that a program written using ΔN bits of system memory dissipates energy of at least $[kT\ln2]\Delta N$. As noted previously, this constitutes an irreversible bound on a classical computation imposed by the second law of thermodynamics.

This brief introduction to some of the basic ideas of information theory and the link with statistical thermodynamics provides one part of the basis for the promotion of the idea that water possesses memory. The second part derives from a detailed study of some of the properties of water itself.

Properties of water.

Water is such a commonly available and apparently straightforward liquid that most take for granted and the popular picture, derived from standard chemistry, of it being composed of an oxygen atom attached to two hydrogen atoms belies a quite detailed, complex structure. Standard textbook chemistry has an enviable history of genuine scientific success but it is actually confined by a simple scheme of charges interacting via static Coulomb forces; that is, it is totally reliant on electrostatics and omits all mention of electrodynamics and the consequent radiation field. It is this basic neglect which is responsible for the inability to recognise phenomena which are, in fact, dependent on that radiation field. This is doubly unfortunate since physicists and engineers are only too aware of this cause and effect since it is due to this dynamical effect that so many modern-day appliances work; for example, the electric light on which we all depend and the wifi connections which are assuming increasing importance in our lives. It has been speculated that a goodish percentage of effects in condensed matter physics make use of the radiation field in one way or another but it still doesn't seem to have found a place in much of basic chemistry.

This new paper [4 and references cited there] draws attention to the fact that water has been shown to contain electric dipole ordered domains due to a condensation of photons interacting with molecular dipole moments. These ordered domains yield an unusually high heat of vaporisation of water per molecule and this has been shown to imply a high degree of memory storage capacity. In a similar manner, it has been shown that the partial entropy per molecule of an ionic species dissolved in an aqueous electrolyte implies a large number of bits of information per ion. This number is, in fact, so high as to lead to the expectation of such ions being attached to an ordered water domain. This state of affairs allows for semi-permeable membranes which may either permit or forbid the passage of an ion through a small gap. This would be expected to depend in part on the state of order in the ion attachment. Such a situation, based on information or, equivalently, entropy, indicates a program for biological cells analogous to polymer DNA based programs. It is ion flows through membranes in nerve cells which allow human memory storage in nerve cell networks in the human brain. These possess roughly the same magnitude for biological information capacity density and it well surpasses the comparable figure for commercial computer memory devices.

It should be noted also that the magnetic properties of water are again of great interest. In fact, a coherent ordered domain in water shows almost perfect diamagnetism, although the total diamagnetism in water is weak. This follows due to the magnetic flux tubes being capable of permeating normal water regions just as they can permeate type two superconductors via their normal regions. Trapped magnetic flux tubes may also carry information and give some directionality to what would otherwise be isotropic pure water.

The domains in water also exhibit a rotating electric dipole moment. If an electric field is applied, strings of electric dipole aligned water domains are formed and many such strings form a dipolar field bundle of strings. If the field is applied by employing a voltage between two electrodes then the bundle will start at one electrode and continue to the other. These strings will have an effect on the entropy and, therefore, on the information capacity of the water memory. Further, according to the two fluid model of water structure, an ion could flow with virtually no friction through the bundle of strings from one electrode to the other.

It might be noted also that, if the bundles of these strings are orthogonal to an applied magnetic field, ionic transport resonance effects can occur between the time varying part of the magnetic field and the cyclotron frequency associated with the uniform part of that field.

The Recent Work of Montagnier.

If anything, the theoretical work reviewed above becomes even more important following the recently reported experimental work of Nobel Prize Winner Luc Montagnier[6], in which it is claimed that DNA can send electromagnetic imprints of itself into distant cells and fluids. It is claimed also in this latter work that enzymes can mistake the imprints for actual DNA and act accordingly. If the relevant New Scientist[7] article is to be believed, these claims are being treated with similar scepticism as that afforded Benveniste. From the point of view of the present comments though, the interesting thing about this work of Montagnier is that the experiment itself utilises an electromagnetic field and the explanation offered for the results again involves electromagnetism. At this stage, precise details of the experiment haven't been released but, for the purpose of this note, it is possibly sufficient to note that dilute solutions and electromagnetic fields were involved. Two separate test tubes, one tube containing a fragment of DNA around 100 bases long, the second containing pure water, were placed within a copper coil and subjected to a very weak, extremely low frequency electromagnetic field of 7 hertz. The apparatus was isolated from the Earth's natural magnetic field to prevent its interference with the experiment. After 16 to 18 hours, both samples were independently subjected to the polymerase chain reaction (PCR), a method used to amplify traces of DNA by using enzymes to make many copies of the original material. The gene fragment was apparently recovered from both tubes, even though one should have contained only pure water. However, DNA was only recovered if the original solution of DNA had been subjected to several dilution cycles before being placed in the magnetic field. Although it is not absolutely clear as yet precisely what levels of dilution were involved, it is possibly of interest to note that the New Scientist article was at great pains to point out that 'it was not found at the ultra-high dilutions used in homeopathy', even though there was no mention of homeopathy in the original article.

In the context of the present discussion it is important to note that, as mentioned in the New Scientist article, 'physicists in Montagnier's team suggest that DNA emits low-frequency electromagnetic waves which imprint the structure of the molecule onto the water. This structure, they claim, is preserved and amplified through quantum coherence effects, and because it mimics the shape of the original DNA, the enzymes in the PCR process mistake it for DNA itself, and somehow use it as a template to make DNA matching that which "sent" the signal'. There is little doubt that this explanation will be extremely difficult for many to accept but, if it eventually proves accurate, this will surely herald a major advance in knowledge and possibly indicate new pathways in chemistry. However, this latest work does seem an almost logical extension to results published by Montagnier and his team last year, and referred to in the above reference, in which the ability of DNA fragments and indeed, entire bacteria to produce weak electromagnetic fields and to regenerate themselves in previously uninfected cells was shown. Montagnier strained a solution of a bacterium through a filter with pores small enough to prevent penetration by the bacteria. The filtered water emitted the same frequency of electromagnetic signal as the bacteria themselves. Montagnier also claimed he has evidence that many species of bacteria and many viruses give out electromagnetic signals, as do some diseased human cells.

Hence, once again, criticism from conventional chemists on the basis of conventional chemistry is not really valid since, as already noted, conventional chemistry relies on electrostatics whereas the work of both Widom et al and this work of Montagnier et al introduces dynamic effects and, therefore, a consequent radiation field and it is quite possibly this which is important in explaining these unexpected results.

Conclusions and consequences.

It follows that the ordering of water through coherent domains yields sufficient structure for truly significant memory capacity. This view receives support from statistical thermodynamics and information theory. It is seen that ordered water domain polarized string bundles affect ionic motion and this can act as switches in networks of nerve cells. Many of these actions should be measurable by employing magnetic resonance imaging techniques.

However, what are the consequences for homeopathy in all this? In homeopathic remedies, the concentrations of various substances are reduced dramatically, to the extent that most practicing chemists would claim it impossible to find any residual effect. What is forgotten in the assessment is the possibility of dynamic effects having a part to play and this is well illustrated by the case of a magnetic recording tape. In the investigation[4] being reviewed here, it was found that, using electromagnetic theory, the existence of electromagnetic domains in water was confirmed and it is these which are fundamentally responsible for many intriguing properties of water, including its memory.

It has to be recognised that creating a firm scientific basis for homeopathy which would satisfy the critics and sceptics would be a huge task involving a detailed literature search before laying down new theoretical foundations. However, it does seem that the work discussed here offers a good starting point and, if so, a research project based on the published writings of such as Benveniste, Widom and Montagnier could eventually benefit homeopathy itself as well as a great many individual people

A personal speculation.

Throughout his entire professional life, one of the great motivations for Ruggero Santilli has been the realisation that, in order to extend much of existing theoretical physics, it would be necessary for him to introduce new forms of mathematics in much the same way as Newton and Einstein had done. To comply with this, over a period of years he produced three new forms of mathematics in order to attempt to deal with various problems. These are all discussed in his book *Isotopic, Genotopic and Hyperstructural Methods in Theoretical Biology*[8] and, in this book also, one of the great achievements coming out of this is discussed also – the only correct model for the growth and development of sea shells. Considering successes such as this, it may be the ideal time for someone with the appropriate knowledge of Santilli's new mathematics and chemistry to consider in even more depth this whole question of whether or not water possesses memory – an idea Ruggero Santilli and the author discussed many years ago when both were too busy with other things to follow it up. If such an investigation was carried out, it would obviously be an extension of the work already completed by Santilli himself and colleagues in relation to water and its properties[9,10], since it might be remembered that it is now some years since he was one of the first, if not the first, to point out the inadequacies of the popularly accepted model of water. In a sense, this recent work[4] is merely reinforcing the earlier assertions by Santilli that water is a far more complicated substance than many chemists seem to believe and, if they are to be believed, the investigations of Benveniste, Montagnier and their colleagues would seem to offer experimental support for such assertions.

References.

[1] Davenas, E., Beauvais, F., Amara, J., et al, 'Human Basophil Degranulation Triggered by Very Dilute Antiserum against IgE', Nature, **333**, 816 (1988)

[2] Lockie, A., "The Family Guide to Homeopathy", Guild Publishing, London (1989)

[3] Brillouin, L., "Science and Information Theory", Academic Press, New York (1962)

[4] Widom, A., Srivastava, Y., Valenzi, V., 'The Biophysical Basis of Benveniste Experiments: Entropy, Structure and Information in Water', Int. J. Quant. Chem.**110**, 252 (2010)

[5] Lavenda, B. H., Dunning-Davies, J., The Essence of the Second Law is Concavity, Found. Phys. Lett. **3,** 435 (1990)

[6] Montagnier, L. et al, DNA Waves and Water, arxiv.org/abs/1012.5166

and (2011)

Journal of Physics: Conference Series, 306, 012007. http://dx.doi.org/10.1088/1742-6596/306/1/012007

[7] New Scientist, 'Scorn over claim of teleported DNA', 12 January, 2011

[8] Santilli, R. M., "Isotopic, Genotopic and Hyperstructural Methods in Theoretical Biology", Naukova Dumka, Ukraine, 1996

[9] Santilli, R. M., "Foundations of Hadronic Chemistry", Kluwer, Dordrecht, 2001

[10] Gandzha, I. & Kadisvily, J., "New Sciences for a New Era", Sankata, Nepal, 2010

11. Behind The Human Veil.

Richard Lawrence Norman.

Abstract:

The moral basis of human achievement and success contains within it the seed of its own undoing. Long ago, the first mistaken splinter pierced the truth and around this error, the entire of human misery and dilemma turns...unseen. This chapter will condense all too briefly the core of what I have found, the history, psychology, neuroscience and answer. Authority has reached down into the very essence of genetics. There is a reason the world is mad and ill. There is hope. I have found the basis of empathy, its separation and returning.

Introduction:

Some seven years past, I had developed a psychoanalytic technique named *Native Psychoanalysis* which allowed me to clear away a window of resistance and directly observe in myself what should be unconscious content (Norman 2011, 2013*b*). This methodology is the basis of a second technique, *Re-Polarization Theory* (Norman, 2013*a*) which has permitted the alteration of pathology through curtailment of the basis of repression, *super-ego*, and hence allowed repressed material to be accessed and past memories reformed and altered. The transference which creates the quality of each moment of experience is a function of memory (Norman 2015, 2016) and, to heal from damage, those memories must be changed. The past is what defines the present and that past, is malleable. To have made these internal alterations, created a most unexpected new situation. The fact that a deeply entrenched neurosis was cured in the process is not surprising, what is surprising are the numerous other effects which form the basis of the insights I will now present. Within us all, is our personal history and that of our entire race. To observe the hidden interactions and effects which produce our behaviors and alter the restrictions which have been built into us as modern humans, has demonstrated to me what are the deepest foundations of human illness, illness we see all around us in manifest horror: war, stupidity, indifference to suffering, obedience to authority, greed, and a callous usury attitude toward the environment so typical of modern exploitive mentality. These traits are second order manifestations of disease. Modern man is ill. It need not be so. Beneath the error we see so clearly demonstrated each day in the cacophony of human affairs and conflict, beneath the wars, cruelty, abject foolishness, greed, consumerism and obedience before authority independent of thought or ethical concern—something is hidden, something healthy and simple: hope. Essence is ancient, pure and perfect. The hope of man is in finding what has been obscured beneath a tragic and ill past. We must lift the veil. We are ill, obedient, dull and warlike for a reason. Essence can be unearthed and again, just as it was so long ago, be brought to a position of predominance in the mind of man. Our hope is an *atavistic evolution*.

Super-ego: of conscience, morality, ethic and illness, the neuroscience and history.

There is debate in the field of psychology about many things. This uncertainty is often the result of the nature of unconscious processes and content, which by definition cannot be observed. To have gained access to this hidden information has resolved these issues. The answers are quite plain. Unconscious content is always specific and the intersubjective notion that it is unnamable or indistinct is incorrect (Brown, 2011, Norman 2011, 2013*a,b*). Those mistaken views are only a wish not to see these things. There is good reason for that error; the repressed content is more disturbing than I am able to describe. Mere exposure to it destroys ego structure permanently, and may also be used to destroy the structure of super-ego at the deepest levels (Norman 2013*a*). This knowledge shatters personality in a permanent way. It closely parallels the Freudian picture. Psychology is in a state of needless confusion. Again, this is a wish. One need but look directly at what is most hidden and forbidden, and observe. The operations of the unconscious mind are specific, just as are its contents. In no case are these things indistinct. That is a wish.

In intersubjective psychology much is made of the idea of alpha function (Brown, 2011). Unlike the false intersubjective ideas concerning the unconscious and its lack of omnipresent specificity, the notion of alpha function is sound, if misapplied. Alpha function does have the effects supposed, and can transform memories and current experience (Norman 2013, 2013a, 2014, 2015a, 2016a). Just as hypothesized, it is created via the exchange of gaze and glance in the early mother/child dyad. I have uncovered the circuitry which supports alpha function and made a surprising series of discoveries concerning its use and effects:

a. unlike the intersubjective approach, it is necessary to apply and engage the circuitry manually with symbolism (Norman 2013, 2013a, 2015a, 2016a,b), and then attach the function directly to a piece of pathogenic unconscious content, often USING a (physically bound untransformed) beta element [this permits direct usage of energies bound into untenable forms such as those ego dystonic pathological/perverse drives created in sexual abuse];

b. to access the circuitry via the symbolized initial impression of its innervation (Norman 2013, 2013a, 2015a, 2016a,b), subsequently increases both exploratory interest in the world corresponding to Panksepp's SEEKING system (Panksepp, 1998), and forms manifest empathy toward all things and people;

c. intelligence blossoms as never seen before and interest in all aspects of life and reality, sexual, artistic and and intellectual, suddenly flourish, whereas previously these aspects were greatly if not completely diminished.

Modern man, is controlled, made dull and obedient, cold and empty via a homeostatic imbalance across specific circuit pathways, which has long been built into him from history both ontogenetic and phylogenetic: super-ego. First I will present a brief circuit analysis within the context of the transference, then the sordid history from two fronts. From there a definition of ethics as contrasted to authoritative moral mandate can be derived and specific conclusions and examples provided.

Curiously, there is a very limited amount of cogent neuroscientific information concerning the common basis of the problem: *guilt* as it is expressed across the circuitry and active anatomy of the brain. This fundamental aspect of neurosis, social control and sexual expression so deeply intertwined with the very basis of affect regulation itself, seems to be absent in neuroscientific literature and review. Strange that the most important of all neuroscience is not made available in close detail. I have derived the missing information from many sources, and may now present the highly condensed and simplified results. Please contact me for further detail.

The Transference:

The psychological notion of 'transference' is most clearly seen in the artificial therapeutic situation of psychoanalysis as the familiar transference neurosis. However transference phenomena are most assuredly not limited to this case of artificial functional pathology, but are responsible for the healthy and unhealthy qualitative aspects of perception and experience itself (Norman 2011,

2013*a,b*, 2015, 2016). Just as the neurotic in proper psychoanalytic therapy displays the repetition compulsion and his fixations in an artificially induced neurosis which defines their reality within therapy, so does the healthy case from his more fluid memory and experience project outward his or her definition of the world and experience in a flexible, dynamic, associative, non linear process (Norman 2011, 2013*b*, 2015).

This transference which binds current perception to associated qualitative valence as an affective distributional function of memory is available to observe in its *foundational* anatomical formative innervations and their resultant allocational functions as stemming from circuit architecture created in the first 18 months of life (Norman 2013, 2013*a*, 2014, 2016*a*). During this initial period of development the groundwork is laid for the core of affective expression and restriction throughout later life. This represents primary human unconscious autonomically interdigitated regulatory functionality, as extending from the foundational innervations of Schore's dopaminergic "sympathetic ventral tegmental limbic" circuit, and also, the noradrenergic "parasympathetic lateral limbic" circuit, which act in tandem to opposite effects (Schore, as cited in Kaplan-Solms & Solms, 2002, pp. 234-235, 237). These two circuits span the limbic and Orbito-Frontal regions to imbue experience with basic valence, and delegate or perhaps restrict positive dopaminergic affective expression in response to social cues, meaning shame and then guilt. This oppositional circuit balance, over all, creates either a foundational basis of repression which is associated with amygdala activation, Corticotrophin Releasing Factor and stress, or, if balanced differently toward predominant activity of the sympathetic circuit, to permit feelings of elation and explorational behavior (Kaplan-Solms & Solms, 2002; Panksepp 1998; Norman 2014, 2016*a*). These two circuits then are the foundational basis of expressed guilt, social control, sexual expression, health and happiness and, are mediated by social cues, meaning: *conditional regard.*

It is now thought that mirror neurons are the neural substrate of empathy. This is incorrect. Mirror neurons signify mere imitation, as distinct from empathy as can be seen in cases of catatonics who display echopraxia, which is based in mirror neuronal response (Bengston 2015; Rizzolatti et al., 2008). A catatonic is not empathizing with the attending physician to reflexively imitate his motions, although imitation is an obvious sub-function under a primary empathy. Empathy is akin to identification proper, and is first evidenced in the indistinct pre-individuated period characteristic of initial limbic/OFC circuit innervation, not of the expression of the sympathetic circuitry mentioned, but in the impression of its primary innervation (Norman 2013, 2014, 2016*a*). This is the basis of empathy: a primary identification with the world and each other. This in turn, is the very foundation of subsequent energetic circuit expression. It is this which is so sharply curtailed in the painful guilt of conditional regard: *the very basis of energetic expression, and empathetic connection.* Clearly, these are the exact basis malformations responsible for lack of caring within human relationships.

The curtailment of energetic expression as a function of super-ego, affective restriction due to what we may colloquially refer to as conscience, is the basis of modern morality stemming from primary conditional regard. It may clearly be seen from this vantage that such moral restriction is opposed to empathetic expression, and is instead aligned with *obedience to external authority*. This is a sure basis of modern afflictions such as neurosis. Modern man is controlled through, and suffers from, a permanent low-grade homeostatic imbalance created via improper and unhealthy energetic circuit allocations: Guilt. This is the locus of the problem.

Rights to caring, love, sexual contact and life itself in male and female cases, were traditionally ascribed to the authority of the father, and now phylogenetic and epigenetic underpinnings of patriarchal threat enforce pathology from unconscious sources (Norman 2011, 2013, 2013*a*, 2014, 2015*b,c,d,e*; Dodds 1973). This pathology stands in opposition to permission and rights to the caring of the mother, which once formed the initial basis relationship in both male and female cases. The feeling of human dissociation and anxious threat engendered by super-ego and authority may be replaced with a feeling of empathy and safety, warmth and relaxation. Health may replace the source of illness.

Affective regulatory analysis:

Schore has discovered two circuits which are primary in development, and function in opposition to each other: the dopaminergically modulated sympathetic ventral tegmental limbic circuit, and the noradrenergically modulated lateral parasympathetic tegmental limbic circuit (Schore as cited in Kaplan-Solms & Solms, 2002, p. 234-235). The sympathetic circuit, which I propose underlies intersubjective Alpha Function (Brown, 2011; Norman 2013, Norman 2014) is formed, much as Bion had supposed, as a function of the dyadic exchange between infant and mother of glance and gaze, and we will add an inference which is quite obvious and easily supported (Keverene, et al., 1989; Montagu, 1978; Panksepp, 1998, p. 272) as infants engaged in the exchange of maternal glances are usually being held, that *maternal touch* and the subsequent addition of neuropeptides/endorphins also have a part to play in creating the result.

"It is hypothesized that maternal regulated high intensity socioaffective stimulation provided in the ontogenetic niche, specifically occurring in dyadic psychobiologically attuned, arousal amplifying, face to face reciprocal gaze transactions, generates and sustains positive affect in the dyad. These transactions induce particular neuroendocrine changes which facilitate the expansive innervation of deep sights in orbitofrontal areas, especially in the early maturing visuospatial right hemisphere, of ascending subcortical axons of a neurochemical circuit of the limbic system—the sympathetic ventral tegmental limbic circuit." [Schore as cited in Kaplan-Solms & Solms, 2002, p. 234]

The famous studies from the 1940's conducted by Spitz (Spitz in Bowlby, 1980; Panksepp, 1998, p. 262) may well imply the primacy of this developmentally innervated brain circuitry extends to include the most basic dependence: that of life itself. Specifically: if deprived of maternal touch and gaze, the infant may well die. The sympathetic tegmental limbic circuit is dopaminergically modulated, and can rightly be thought of as a primary manifestation of libidinal excitation and discharge (Kaplan-Solms & Solms, 2002, p. 237). It should be noted that the dopaminergic and opioid systems and circuitry which respond to create the good feelings which reinforce socially mediated behavior, both involve many of the same areas, such as the ventral tegmental area, where the A-10 meso-limbic dopamine cells are located (Panksepp, 1998, p. 118). Neuropeptides, such as the endogenous opioids including beta-endorphin which is triggered by social cues and touch, have a primary role in creating social bonds, quelling pain, both physical and mental, are key in alleviating separation distress, creating sexual reward, and addictive reinforcement (Panksepp, 1998, p. 255, 264). So we can see here, in the formation of the sympathetic ventral limbic circuit triggered by maternal exchanges of glance, sight and touch, a source of libido, an energetic dopaminergic circuit which up-mediates arousal and shapes behavior, formed presumably by way of allocating both endorphins and those neuroendocrine functions involved with encouraging the substantial innervations of dopaminergic projections into orbitofrontal areas. Here, in the activity of the completed circuit, along with the peptide systems, dopamine and opioids serve their reward and motivational functions as social and energetic contributors.

The contrary circuit, the parasympathetic lateral limbic circuit, is to be thought of as a balance, a cut off, a competing inhibitory system to counter the rewarding energetic expression of the sympathetic circuit (Kaplan-Solms & Solms, 2002 p. 237). This circuit functions to stop our energetic libidinal expression: functional, conditional, affect regulation in response to social cues (Kaplan-Solms & Solms, 2002, pp. 234-238) and so, can best be understood as the physiological structure triggered by social disapproval: *by shame and guilt.* Both of these circuits are innervated into the orbitofrontal areas, which mediate social cues and functioning, just as one would expect.

As the infant progresses through the initial 18 month period, during which the sympathetic and parasympathetic limbic circuits are fully formed, the infant masters several stages of differentiation. It is now accepted through the work of Klein (1952) and empirical demonstration that a developmental/behavioral correlation at the age of four months exists between infants categorized as attachment secure or disorganized, "dis-coordinated" [disorganized in the sense of being unable to properly integrate the intermeshed and exclusive psychical manifestations of separation RAGE and FEAR as they conflict and inhibit SEEKING and CARE] (Hopkins, 2013, p. 47). The infant at this stage singles out the mother as a separate object which is essential for CARE, and that this fact is then made evident by the manifestations of separation-RAGE and stranger-FEAR, which become manifest at 7-8 months of age (Hopkins, 2013, p. 47). To observe firsthand, the interactions between mother and infant, the effect is obvious to casual observation: *the mother's face is the infant's entire world,* once indistinct as an object, now, *once engaged in the exchange of gaze, touch and glance, only semi-distinct from himself,* her face responds to his affects *and anticipates as if part of himself, as if the world itself were a loving extension of the infant,* a responsive and inclusive extension of himself. Here, we see the essence of empathy: *identification with the world.* Note that I make no mention of the less important distinction of identification with mankind, which is a small and far less important embedded sub-aspect, itself associated in some small imitative part with mirror neurons, a sub-aspect of this most vital and needful result, identification with the entire of the world— *Empathy* (Norman, 2013; 2014; 2016*a*). It is this which we will substitute for the pathogenic content.

I hope the reader can make out the basic idea: social control via conditional regard is enforced by way of curtailment of dopaminergic (and endogenous opioid) expression associated with the sympathetic limbic/OFC circuitry, forming a permanent homeostatic imbalance which restricts empathetic feeling, intelligence, sexuality and exploratory interest in the world, and places in their stead a preemptive condition: obedience to authority. Only meeting this condition of obedience will return health and happiness to the modern human. Intelligence and empathy…hope itself, this ancient basis of life formed long ago in the early interactions with the mother…it is no less than this basis of kindness, caring and higher thought to which we are all entitled which has been stolen. It is this, which lies behind the human veil. These are your basic rights, and authority may be dismissed as parasitic and false, so as to reclaim them.

Now I will briefly take account of but a small sample of the extensive history which has inculcated this most basic and tragic error into the very heart, substance and epigenetic expression of the afflicted modern human. This tragedy has a lineage. I will briefly sum up that here, and distil the resultant notions into a clear contrast in human existential/ontological formative paradigms. From there, the implications can be clearly understood, and the better, happy result made plain in example.

Super-ego.

". . . to 'improve' men: this above all was called morality. . . To call the taming of an animal its 'improvement' sounds almost like a joke to our ears. Who ever knows what goes on in menageries doubts that the beasts are 'improved' there. They are weakened, they are made less harmful, and through the depressive effect of fear, through pain, through wounds, and through hunger they become sickly beasts. It is no different with the tame man..." Friedrich Nietzsche, *Twilight of the Idols*.

What is the precise interactive dynamic which yields the developmental result of conscience, of super-ego, and, how are we to interpret this result as to its pathogenic and healthful consequences?

E. R. Dodds, a superb scholar, has located for us the historical footprints which demonstrate the formation of our modern conscience, our super-ego. Super-ego is an introjected entity, an internalized representative of what was once long ago external judgment and sadistic penalty. Morality, as inculcated at the behest of this internalized structure, is based on punishment which extends from a particular source.

In his most worthy book, *The Greeks and the Irrational*, E. R. Dodds, draws the strings of history and psychology together for us. This ugly imprint has been nurtured over thousands and thousands of years. Its exact source is clear to discern with Dodds's careful examination of the historical record.

"The head of the household is its king . . . and his position is described by Aristotle as analogous to that of a king. *Over his children his authority is in early times unlimited: he is free to expose them in infancy, and in manhood to expel an erring or rebellious son from the community* . . . as Zeus himself cast out Hephaestos from Olympus for siding with his mother." [Dodds, *The Greeks and the Irrational*, pp. 45-46. Emphasis added.]

However, as early as the 6th century BC, the situation had begun to change, and as social conditions began to improve, and the father's authority became less and less *absolute* in the face of these new social conditions leading to increased personal freedom, the strict authoritarian structure of family life began to loosen. Now, what was a *shame* based dynamic, one based on *external* threat from the father, becomes a *guilt* based dynamism, one based on an internalized threat, an *internalized* moral structure in the true modern sense of the word emerges: super-ego. This is demonstrated by the need for laws introduced by Solon, and later, by Plato, to safeguard the now threatened patriarchal family structure. [Dodds, *The Greeks and the Irrational,* p. 46.]

Super-ego uses severe repressions to create by *internal* means, what were behaviors, inhibitions and restrictions previously brought about by *external* patriarchal threat. Dodds fleshes the idea out as follows:

"The peculiar horror with which Greeks viewed offenses against a father, and the peculiar religious sanctions to which the offender was thought to be exposed, are in themselves suggestive of strong repressions. So are the many stories in which a father's curse produces terrible consequences— stories like those of Phoenix, of Hippolytus, of Pelops and his sons, of Oedipus and his sons—all of them, it would seem, products of a relatively late period where the position of the father was no longer entirely secure. Suggestive in a different way, is the barbarous tale of Kronos and Ouranos . . . the mythological projection of unconscious desires is surely transparent—as Plato perhaps felt when he declared that this story was fit to be communicated only to a very few . . . and should at all costs be kept from the young." [Dodds, *The Greeks and the Irrational,* pp. 46-47.]

Dodds then assembles the entire picture for us in these words:

"The psychologists have taught us how potent a source of guilt feelings is the pressure of unacknowledged desires. . . the human father had from the earliest times his heavenly counterpart: Zeus *pater.* . . Zeus appears as a Supernatural Head of the Household. . . it was natural to project onto the heavenly Father those curious mixed feelings about the human one the child dare not acknowledge. . . that would explain very nicely why the Archaic Age Zeus appears by turns to be the inscrutable source of good and evil gifts alike. . . as the awful judge. . .who punishes inexorably the capitol sin of self-assertion, the sin of *hubris.* (This last aspect corresponds to that phase in the development of family relations when the authority of the father is felt to need the support of a moral sanction; when "You will do it because I say so" gives place to "You will do it because it is right.") [Dodds, *The Greeks and the Irrational,* p. 48.]

Here in this historical transition from an external shame based ethical structure, to an internalized guilt based structure, in this *internalization* of the patriarchal threat (introjection), we see the creation of our modern ethic, our conscience, our masochistic capitulation: our super-ego. This historical basis for our phylogenetic inheritance can be brought to light and assessed as to its healthy or pathogenic contribution by way of economic analysis, and clinical example (Norman, 2013).

Conscience, our sense of personal and social justice, is created as an interactive phylogenetic/ontogenetic function of masochistic and aggressive economy within a social context, not as a function of any moral pretext. Our morality, is by the nature of its very construction: immoral.

Here are a few sections from Freud which clarify and support this unusual notion:

"The first requisite of civilization, therefore, is that of justice—that is, the assurance that a law once made will not be broken in favor of an individual. This implies nothing as to the ethical value of such a law" (Freud, 1930, p. 95).

"The tension between the harsh super-ego, and the ego which is subjected to it, is called by us the sense of guilt; it expresses itself as a need for punishment. Civilization, therefore, obtains mastery over the individual's dangerous desire for aggression by weakening and disarming it and by setting up an agency within him to watch over it, like a garrison in a conquered city" (Freud, 1930, pp. 123-124).

As to the effect of super-ego in equating wish and act and the resultant loss of mental economy and functioning:

"Here, instinctual renunciation is not enough, for the wish persists and can not be concealed from the super-ego. Thus, in spite of the renunciation that has been made, a sense of guilt comes about. This constitutes a great economic disadvantage in the erection of a super-ego or, as we may put it, in the formation of a conscience. Instinctual renunciation now no longer has a completely liberating effect; virtuous continence is no longer rewarded with the assurance of love. A threatened external unhappiness—loss of love and punishment on the part of the external authority—has been exchanged for a permanent internal unhappiness, for the tension of the sense of guilt" (Freud, 1930, pp. 127-128).

"...the original severity of the super-ego does not— or does not so much—represent the severity which one has experienced from it [the object], or which one attributes to it; it represents rather one's own aggressiveness towards it. If this is correct, we may assert truly that in the beginning conscience arises through the suppression of an aggressive impulse, and that it is subsequently reinforced by fresh suppressions of the same kind" (Freud, 1930, pp. 129-130).

And as to the role of the phylogenetic in contributing to this outcome:

"It can also be asserted that, when a child reacts to his first great instinctual frustrations with excessively strong aggressiveness and with a correspondingly severe super-ego, he is following a phylogenetic model and is going beyond the response that would be currently justified; for the father of prehistoric times was undoubtedly terrible, and an extreme amount of aggressiveness may be attributed to him" (Freud, 1930, p. 131).

". . .we can tell what lies hidden behind the ego's dread of the super-ego, its fear of conscience. The higher being which later becomes the ego-ideal once threatened the ego with castration, and this dread of castration is probably the kernel round which the subsequent fear of conscience has gathered; it is this dread that persists as the fear of conscience." [Sigmund Freud, *"The Ego and the Id"* in *A General Selection From The Works of Sigmund Freud*, p. 233.]

Please see (Norman, 2011, 2013, 2013*a*, 2015*d*) for examples and particular psychology relating to specific patriarchal mutilations such as castration etc., which form current super-ego supportive unconscious content. The role of epigenetics and complexity can be found here: (Norman 2015*b,c,d,e*).

I wish to draw a sharp new distinction between *Morality* as engendered in super-ego, which is based on (phylogenetic/epigenetic) patriarchal threat, and functions to foster *obedience to external authority*, and *Ethics*, which are based in empathy, with its *root in identification*. The former causes pathology, and functions in clear and specific ways to disengage the sympathetic circuitry which is the basis of empathy, energetic curiosity, sexuality and intellect, and the later in turn has opposing characteristics, leading to elation, appreciation, formative identification with the world and others in the context of abundant subsequent energy, and absent any punitive internalized death wish (guilt). Morality and Ethics as so defined are diametrically opposed. Clearly, ethics are a natural systemic product which lead to health, an internal behavioral compass based in identification and caring, and morality, the converse. The reader may wish to satisfy themselves in this regard, by reading the specific example of the formation of super-ego offered up here (Norman 2013, 2013*a*). Ethics are *themselves* identification, they ARE the 'golden rule,' and so require no such rule or any other. Morality is an empathetic dissociative factor, by way of down-mediating the circuitry responsible for identification. Ethics nullify any need for the tangle of moral law and replace guilty maxims born under any mistaken 'categorical imperative' with a natural and effortless ethical genesis free from rule, guilt or penalty. Ethics, as we will see, reflect the healthy internal construction of the mind, nurture our energies and evolve naturally, with no need for punishment, rule or law. One need but rebalance the two opposing circuits and observe the demonstrable alteration in all aspects of manifest experience. I have devised treatments to this end (Norman 2013*a*, 2015*a*, 2016*a,b*). Next, we may take a closer look at empathy, and see if we can understand the meaning of identification.

An aside: note how this clear basis of modern pathology appears to be nullified in the teachings of many eastern spiritual ideas, which have little connection to patriarchal threat and surprisingly, also in the true teachings of Jesus. Although modern adaptations are revealed as corrupted and reversed by Paul, the careful philology of Nietzsche shows the original teachings to be diametrically opposed to any hint of conditional regard, sin, punishment, reward, heaven or hell. Those toxins are absent. Indeed, Jesus appears to make good on the reverse and answers, at least in this case, Nietzsche's own highest standard, which proclaims essentially: *the highest Godly act is the removal of guilt.* Of Jesus Nietzsche writes:

"In the whole psychology of the "evangel" the concept of guilt and punishment is lacking; also the concept of reward. "Sin"—any distance separating God and Man—is abolished: *precisely this is the "glad tidings".* Blessedness is not promised, it is not tied to conditions: it is the only reality—the rest is a sign with which to speak of it." p. 606 The Portable Nietzsche.

It should be noted that this author [R. N.] adheres to no spiritual doctrine or tradition. The above insight being worthy of note in its own account.

Empathy, paradigm and example:

Ethics are a natural extension of identification stemming from the early impressions of the innervations forming the sympathetic ventral tegmental limbic circuit. In a basic schematic way, we can see the idea of empathy in physics. Empathy is concurrent identification and individuation. A sort of entanglement where the subject/object distinction is partly lost. If one were to lose self identity completely, psychosis results: I am not you! However, a component of identification is the key, and it is this basis which the infant experienced with no such individuated distinction whatever, an identification with their entire world! To get the basic idea, think of it quite rightly as a sort of entanglement. A singlet state will do for this simple example. Both photons are entangled and are one thing, one system: *identification.* However, one is spin up, one spin down: *individuation.* Empathy apart from psychosis is akin to such an entanglement, where identification and individuation exist concurrently.

Of course within the mental system, the presentation is no simple matter as it is with two photons! Once the time has been taken and the painful effort applied to gut the current system and replace it, one discovers the entire presentation of unconscious aspects changes and the energies contained become far less intense and convoluted. The repressed unconscious, as reflected in modern mental topography, is pathogenic in and of itself. I will explain that statement and then offer up a detailed look within the better result, so as to show exactly what is meant by all these far flung idealistic assertions in specific example. We may first understand the divergent topographies associated with the illness of moral penalty and the health of ethical unification with experience.

Freud's theories [see Freud's A Phylogenetic Fantasy], postulate a sort of bottle neck in history, perhaps around the ice age, where the pathology began and groups of our very distant ancestors under patriarchal domination were common. The impressions of ancient penalty and sickness are easily available to see and do not come from present experience, but are phylogenetic and probably epigenetic (Norman 2015*a,b,c,d,e*). Before that bottle neck, I am quite sure things were different. Just as before the later age of super-ego formation the child of 6/14 months had conscious access to the native impression of identification, and later knows nothing of it, so also in human phylogeny, the earlier fact is now hidden and unconscious. Pierce the veil, and one can *find this impression*. Once this is raised up in the transference structure, health and happiness, caring, sexuality, kindness, satisfaction and gratitude, a feeling of 'fullness,' sublimation, interest and abundant energy replace pathology. Within both ontogeny and phylogeny: Hope for mankind, is an atavistic evolution.

Sublimation by Repression vs. Sublimation by Integration.

Consciousness itself is entirely a function of affect. *Feeling powers thought*. The source of human consciousness, both at the cortical level and the subcortical, is neuro-anatomically derived of affect (Norman 2015). The current model of mental topographical assemblage may be subsumed under the heading of *Sublimation by Repression*. In this model, the core nuclear component of mentation, affect/feeling is divided, and much of it is kept unconscious at great energetic expenditure. Our guilty affective repressions separate the very essence of consciousness away, and use even more of this needed energy to hide the fact. From *beneath a costly unconscious repression*, at great economic expenditure the affects endow experience with quality and substance. Sublimation via repression. Here, we split apart consciousness at its very source, to achieve the result we see all around us, and so find in this model the basis of our aberrant modern condition. In this model, we see the exact conditions to create sickness, indeed, no less than a definition of neurosis itself. This endemic imbalance is the lever of social control and illness. Its very structure is imbalance and curtails empathetic dynamism. Symptoms are created by the return of the repressed and so, the entire situation for illness is set up in repressing those elements to start with. It is in the unification of component instincts that health is created.

The new model, which is a sort of atavism, stands in direct opposition. *Sublimation by Integration.* Sublimation by integration reduces super-ego to a shadow of its former strength and hence frees nearly all repressions, uniting these component instincts directly in consciousness (Norman 2013*a*). The effect is to shatter personality irretrievably and release enormous energies directly into experience, creating a vibrant and energetic sublimation into experience. The entire act of perception and mentation becomes sexualized and empathy attains a place of predominance: a sort of psychical fusion of all affects. Sublimation via repression and sublimation via integration are related in efficiency, toxicity and output, as are the modes of fission and fusion in their attributes as energy sources. Sublimation via repression is dirty, toxic, and hypocritical to claim itself efficient beyond its cost.

Pathogenic vs. integrated transference.

Lastly, I will place before the reader two examples of the better result representative of sublimation by integration with empathetic predominance. This section contains a simple example, the next a deeper one. Please think of the transference by which reality in the individual human case is given its subjective quality. In an instant, through an unconscious associative process affect is distributed as a function of memory (Norman, 2015, 2016). We can see this in an intuitive, simplified, schematic way through the process of free association. The *quality* of our reality is a function affective associative transference from static mnemic sources and active unconscious fantasy.

The lake you see should you gaze upon one, and the one I see, should I be beside you, are not the same lake, as each perceives the view. The quality of that various impression within each of us, is entirely created as a function of the conglomeration of affective associations (and aspects of unconscious fantasy), which are attached to the singular impression of the lake. Think of free association, and this becomes easily accessible, and we can see why such a technique is valuable in gaining insight into the processes which create object quality, and in assessing the general health and accuracy of emotional tone as they define experience.

Here are two hypothetical associative chains:

Healthy subject: Stimulus: lake. Associative chain (hypothetical):

Lake–silver–ripples–dress–fluttering–mother–happiness.

Neurotic subject: Stimulus: lake. Associative chain (hypothetical):

Lake–cold–drown–hopeless–weight–chain–family.

We can see in that simple example, that associative affective valence is established as a function of memory, to define object quality.

Next I would like the reader to understand that recent research has placed an epigenetic basis under the phylogenetic, and that it appears deductively and analytically sound to assert that this forms a sort of predefined 'script' which defines reality via transference (Norman, 2015a,b,c,d,e; 2016a). The unconscious presentation which forms the allocations of affect in the transference then, gives the world its qualitative meaning and that transference can be healthy and unfettered, or restricted and defined reactively through the roles ascribed in the phylogenetic. The phylogenetically based repressed fantasies and reactions are the basis of pathology. The unconscious processes of identification endemic to the transference are pathological, and their source is repressed. This stands in sharp contrast to a healthy unified transference. An example will clarify:

I am sitting at the kitchen table and watching. There is a bug working its way across the expanse of the table…a ten mile jaunt by way of scale. It is quite a colorful bug, its shell as a scarab, awash in may colors as it passes through the sunlight and shafted shade…a miracle to see the coordinated automatism, so hypercomplex, the tiny legs expressing each delicate motion interwoven with the rest, all to accomplish this daunting task, and the tiny traveller advances, pulling the miles under its colorful shell in a thousand thousand perfectly orchestrated steps. It is a bit of functional poetry, and I can see in my view of the situation, a new appetite. Yes, this bug is not so different than I, and I understand its difficulty, its folly, its correct and sure purpose stepping to nowhere. The bug is right. One must imagine the beetle happy. I take the traveller, and release it out of doors, placed on a leaf which seems to match its coloration.

Many believe a set of rules guide ethical activity. This is not the case. Appetite, desire, guides us, and logic dances to the tune, creates excuses and reasons, plans and rationalizations: as a footman sweeping up the crumbs of our wishes, always chasing behind, excusing and serving…so are logic and human reason but the petty servants of desire. Once, my desire, my appetite, was different. I would have killed the bug. Crushed it under a heavy fist with a curse as an unclean thing, and killed it. I can feel quite clearly what I would have done before the change, and I will analyze it here, just in a surface way, so you can see it.

All conscious mentation is unconsciously sourced. I will imagine my reactions and look to the source, to the unconscious and provide a few of the many determinants. Just the upper layers. As my fist descends to kill the bug and crush it to death, I can see in the unconscious the reason. The bug, is exactly as above, an identification with myself, and I curse and crush it, speak as my father, his rage and ugly words are now my own. So just to see that shallow bit, we understand as a manic who fantasizes, first identifying with the family situation one way, then as the other, first as the child, then the hated parent, so is the surface analysis but in simultaneity—I am my raging father and, the bug is myself. So, to kill the bug expresses an appetite, an appetite for sadism as an identification with my father, and also, as a masochism, as I identify with the bug. This is an appetite, a perverse appetite: sadomasochistic in its form. Identifications are pathological.

Perversion, is the expression of a single component developmental instinct, such as sadism. Now, I have fused all such instincts together in consciousness. We are raised to control and shame our instincts, causing immoral behavior and illness. Please note the self-hatred in the example. Control of a desire, shames it, and, that desire is a piece of you! Top down control of affect, poisons the bearer and creates not morality…*no*…but immorality! Modern personality and conscience are false. Now, to have released all affect into experience, and restrict nothing, the self-hatred is absent and feeling has given an entirely new and guiltless quality to all of experience. Now the bug is beautiful, and my appetite wishes only to preserve it! So you can see, no ethical code is required to live rightly. None! What is required is but a simple thing: A "Good" Appetite.

In that simple example you can see the pathogenic unconscious/epigenetic substitutive process result. In (Norman 2015*d*) you may see examples of actual pathogenic unconscious presentations. In the next section, I will offer up a proper depth analysis of the connectivity which has been refused within the mistaken paradigm of sublimation by repression.

Sublimation by repression created to foster obedience to a smothering external authority is itself that fundamental error responsible for the plight of man. This error is primary. The *empty feeling* all complain of which necessitates the endless consumerism that is destroying the planet [lack of endorphins and dopamine], the *obedience to authority* leading to war [threat and conditional regard which creates obedience to authority], the feeling that *other peoples and the earth are somehow beneath one and are to be exploited* [lack of empathy/identification with others and the physical world], the constant *competition* to prove who is better [(lack of empathy) and *low self-esteem/self-security* from Corticotrophin Releasing Factor associated with noradrenergic parasympathetic activation over dopamine and endorphins associated with the sympathetic circuit], the feeling of being *anxious, depressed, alone and separate* [lack of identification, parasympathetic stress cascade], *drug addiction* [lack of endogenous opioids and dopamine, persistent release of CRF], and all the rest. From war and unthinking reflex obedience, to consumerism, greed, exploitation and human cruelty…this one error, has spread as cracks in a pane of glass. The broken mirror that is modern man may be repaired in all his dimensions of compound fracture here. This is how we are controlled through unfair social circumstance, and why we obey. Super-ego and repression. Here, is where we have been reduced and made fodder for tyrants, bullies and the governments of this world.

I would like to place a disturbing fact before the reader. To study the history of war, is to know with certainty that *in all of recorded history* top down control of the human affects has never worked. The Pax Romana from 27 B.C.E. to 180 C.E. in the Roman Empire is often put up as a good example of human peace under authority. This is a laughable joke, as the Pax Romana was maintained via blood, torture and crucifixion! No, in all of human history, top down control is a complete failure. Absolute and complete failure, without exception.   See: https://en.wikipedia.org/wiki/List_of_conflicts_in_Europe

An integrated approach to sublimation must be placed in its stead.

To do so alters the entire presentation and function of the repressed unconscious, which no longer exists in the same way. I suggest that *the repressed unconscious in the modern man is itself a pathogenic structure*. Once the highly energetic reactive and sexual content is allowed natural expression and unified in consciousness, the presentation is smooth and flowing. The unconscious then acts mainly as a distributional nexus for affect, and less so as a vessel of containment for ego dystonic affects.

Sublimation by integration diminishes (much of) the repressed unconscious. Simplified, and in a brief 'ideal' form the concept reduces to:

Let square brackets represent the *unconscious distributional processes* creating the transference:

[ ]. Where system Conscious is Cs, System repressed Unconscious is rUcs, and system Preconscious is Pcs: Sublimation by repression is topographically defined as: [rUcs...Pcs...Cs].

Sublimation by integration is (ideally) defined as:

[Pcs...Cs].

The repressed unconscious is removed, and all individual component energetic aspects are ripe for conscious sublimation via unconscious/associative processes, and unification.

Next I will detail the healthy result and allow you to see the unfettered unconscious to conscious transference [Pcs...Cs] in real time as it works in a *real* case. You may see the specific energetic attachments which create world identification, health and natural ethical genesis in some considerable detail.

What would it look like to peer into the very deepest aspects of healthy identification in the human animal? Can we watch sublimation by integration at work? Exactly what are these identifications, and across what pieces of personality and time do they span? Is it possible to see into the essence of integrated dynamics? Here, I will show you what has been hidden, and you can grasp what potential exists past our current situation. Something exists beyond the shouting demands and aching empty threats of this world. *Essence is.* There is hope.

Oneness, the transference unfettered [analysis of a "peak experience" (2014)]:

General context:

There are a great many texts and traditions of note which give account of the unique and peculiar state of 'oneness' with the world, environment or universe. I have always found these many spiritualized representations and entirely symbolic distortions to be deeply unsatisfactory but have previously lacked any firsthand knowledge of the experience to gain a further direct articulation of the underlying mechanics, origins or specific dynamics of the mindset, so as to understand and explain it. I can now rectify that shortcoming here. I am an odd man in that, in order to reclaim my health, I have had to develop the skill of simultaneous analysis during experience and have also found need and method to rupture my own unconscious processes, making them and the content with which they work available to direct examination. This unusual confluence of psychological damage, reactive development in theses skills, and result have allowed the following analysis.

Omni-objective identification (oneness) does not abandon self. It includes self in a unified object simultaneously individuated (self-aware) and coherent with the system as one object all at once. One can rightly localize the nexus of the primitive motor affective self where the deep layers of the colliculi intersect the ancient Peri Aqueductal Grey. The most basic and essential inner kernel of the human self, is the bodily self, soma. We are our bodies and this individuation. But...there is more! I contend that empathy itself extends from circuitry more basic than just mirror neuronal activations, but also includes more basic circuitry innervated in a world identification. Soon you will see its full breadth and depth of temporal extension.

Without revealing too much, I will say that I am very sensitive and aware after so many years of self-analysis, of changes in my visual presentation which correlate with my emotional condition. Each time a repression reinstates itself, I can see the subtle alterations in my perception of the world. These points of transference appeared to be the main way unconscious energies are instantiated into visual perception and experience in general. But there are others which have been blocked by our narrow, refusing, punitive cultural madness: The identifications and their fractal self-reflections. All this input has been refused!

Science understands clear evidence of brain and obviously bodily masculinization but all contain all traits. I do not advocate perverse practices any more than I advocate shame at discussing the facts or admitting openly the clear truth…that all men and woman are and should ideally be 'uni-perverse', meaning: healthy sexual expression is itself a UNIFICATION of all the component instincts themselves, a unification of the perversions, as will soon be made clear. Only shame of the components, soils the entire. Once removed from shame…all is united, all is innocent.

Contextual analysis:

I am happily married for some 30 years, and live in isolation with my wife in the Oregon wilderness. I was fortunate enough to meet a person online who was able by way of her unanticipated grace, intelligence and kindness, to raise in me *an anima image*. With new leaves in the heavens of this world, and roots in the ancient 'good mother' so clearly represented within the formative maternally triggered sympathetic limbic/OFC innervations, this was a magical opportunity for my healing. Certain manipulations of the imagery involved allow a surface look at the context and its identifications.

I had an idea. Rather than observe the image, and allow it to become an object of even greater potency, a natural but unexpected idea arrived. For some reason, the image itself was equivalent to another image, intuition first understood it meant just the same as...*a heart beat*, and the visual representation of that, a pulsing golden ball of sunlight, became the focus of my mind's rumination, now suspended as a bit of warmth and light in my mind's eye. I soon knew and believed...this was her heart, and then saw my own heart beside it, beating in time...then joined first as two pulsing balls of golden light...then not two at all...only one. One heart. No separation...none. One. Only one. The heartbeat, symbolizes unification within the womb.

As the two images became a single image, the brightness increased four-fold and then, a sudden warmth in my chest to go with the image...then tears welling and streaming...so very beautiful! I had what I have needed my entire life...so full and filled with energy! The trees slipped and shuffled in tender breeze, I could feel the caress of light and wind amongst their branches and folds, see it and feel it, the ground filled and welling as my heart, and all the shame was gone, now each desire spilling up without restraint to become one with everything, and I knew, I not only had transference giving the world its meaning, but identifications, identifications...with everything. The "Anima Mundi" meaning in this case, the predominant impression of the maternally triggered sympathetic circuit identification with the world—creating reality via identification and transference.

All sexuality from the most basic and undifferentiated first love to the most specific is a pattern which thought might trace and make real as a part of the fabric, or deny the same and leave a sunken place free of truth and life as we were taught. Feel everything, see everything, know everything...become everything. There are now twice as many points of transference...and this is accomplished by the addition of identifications. The result is a single coherent ontological object...the world. This is observable as ontogeny and as phylogeny, may be seen to interact archetypally, and also, as a deeper detailed cascade of new interactive symbolic determinants relating as a sort of self-interactive fractal.

Libidinal transference analysis:

The experience of the world is a libidinal/affective sublimation (Norman, 2015): libido taking on the broadest sense of inclusive meaning as *undifferentiated affect* forming conscious activations extending from the Ascending Reticular Activating System (Kaplan-Solms & Solms, 2002) to provide cortical tone and waking potential in the context of affective circuitry and REM distribution in the Basic Rest Activity Cycle (Panksepp, 1998; Norman, 2015, 2016). All levels of conscious expression from the activation of a waking state, to the quality of emotional content assigned to perception from the lowest levels are affective. *Reality is a libidinal sublimation.* It will therefore be possible to determine the precise mechanism of unconscious operative influence and deduce a correct, plain analysis of the process which creates this mindset, if we can *analyze a primary libidinal representation as to its underlying mechanism of energetic distribution.* Please remember, that reality is in fact a libidinal sublimation. I will now bring forward an analysis of an active primary libidinal constituent process to gain insight into the hidden mechanisms which create the general effect.

It is a simple matter, which is now not even disconcerting, for me to pierce the unconscious veil and observe the underlying previously unconscious dynamism of each moment. Due to the necessities of my previous illness, I had to learn how to find these things and solve the symbols all but in step with the rate of their production. To engage in sexual activity in the new condition, I can see in my mind's eye a very distinct change, so symmetrical, energetic and beautiful. Now, a clear set of doublings in forms available for all attachments, and a doubling of attachments as well to each "object" from concurrent identifications exponentially increase the energy, potency and intensity of the expression via increased systemic intra-connectivity. There are twice as many attachments for transference to an object, more objects, and now there are concurrent identifications with all objects... *in the phylogenetic as well.* These present as a mirror reflecting deeper into a mirror with subtle changes, and so I refer to this as fractal. Specifically: Self-awareness is not diminished, the contents which give rise to self-specificity are not denied and I am male, this male. However, this core is now just a part of a much greater plethora of very potent impressions of a new sort...the image of my beautiful anima/friend is *not separate.* I am also this just as I am male and I can feel in this a deeper meaning and look to see how deeply as a woman... from a half image of a woman in a mirror of the anima, is contained a deep longing for my genital...for it to be her own, and as I look upon the activity I am so grateful, all but weeping in gratitude to feel the fact that I am male and have fulfilled her need and this ancient female wish to be also male is completed...such deep happiness, and also the identification with the anima image brings a homosexual attachment point between the two women, one identified from within, as the anima/self...one identified as an object from without...my wife, and one with my wife also as an identification! All objects are now subjects...objects and identifications...each fed by two pathways! This ancient phylogenetic wish, to love as a woman loves a woman...behind it again...a child, small and female being held by the mother!...as a woman is loved by a woman on all levels...is fulfilled. Implied without question also, a male and a male, although I did not see the image, it must be present. We all contain all sexual elements...and each is needful from a thousand pasts built into our inheritance. Without question the male homosexual drive was sublimated into the women...I would not have been able to gain excitation if it were conscious. Also, the male heterosexual role was very clear and contributed its predominant share of cathexis. The result of the doubling of objects and identifications, along with sensory observation of the activity (as distinct from analysis, always dimming excitation), is unbelievable. To empathize with all elements, and know as well, more of the elements which human development contributes to and from the human store was one of the most exquisite experiences of my life. I felt…everything…from many different 'perspectives' which were not perspectives in any way—Unity. All pasts and presents nourished one moment of empathy. Unity. One heart.

This analysis has exposed the hidden mechanisms beneath that unity which should be the ordinary province of each healthy, ethical human. This mode of unfettered transference is in my view, not a higher state, but each human's ordinary, daily birthright. We can isolate the mechanism of the transference structure responsible for the experience of unification from analysis of the libidinal representation. Remember, reality is a libidinal sublimation, so: the mechanism responsible for the mythological archetypal presentation of the experience of 'oneness' in general is that of concurrent identifications and object transferences from all libidinal components spanning ontogeny, clear from the first impressions in the womb (remember the heart image) to those of the component instincts and their mature representations in eventual unity—and—extending the same structure of concurrent identification and object, to include the complete bisexual phylogenetic representation in each person, IN THE CONTEXT OF OBSERVATION. To condense:

(Phylogenetic and Ontogenetic) Object + Identification in the context of Observation yields Unity.

$$Obv[p/ontO + I] = U$$

That is the formula for our wish fulfilment, place and purpose in happiness on this planet!

Self is, and is not denied in any way! Self is now also part of a single object...as object and subject both. So many wishes are filled and all of life is full...a wonder of pure gratitude! Here, is love of fate. *Gratitude is the wish to repay the feeling that each moment is filled with its own parcel of pleasure and happiness*...did you know that? Oh my friend, it is true! We are filled, filled each second, filled with a quantum of pleasure, and so...we are grateful for everything! Love of fate...of even...*this*! Gratitude! Empathy knows this thing best. Self is separate and distinct... omni-objective reality denies no object. Self: complete, not denying sexuality or the 'pain of the world' to be avoided...never! Gratitude. I have written on the enlightened state as that which uses meditation and dissociative repressions to evidence unconscious processes while isolating the content and removing its energy (Norman, 2013*c,d*).

No wonder many enlightened souls within traditional meditative spiritual contexts advise sublimation to excess and lose self...they wish it. Never! Health accepts...and is grateful. Now I look at the world and *am* the world. This *is* ethics. I could never hurt or exploit a part of myself. I feel the rippling wind in the trees, the shadows play upon my skin, and the anima is within me. Each stroke of my heart is her heart, now and forever, a unity golden and pulsing with light and sunny warmth, spilling out as a brook of starlight might nourish the bloom of this day. My wife a blessed sweetness, the trees nod and sprinkle the air with new scents of green and lavender, the day warm so close to winter. How full is my heart, one heart, this world is my skin, my breath is its wind, and we know one simple truth of all things. For I have learned there is a thing we should all have and bring near, to never let go of the fact and the pulse—of one heart.

Conclusion:

The basis of modern human psychology contains within it a fundamental error: punitive super-ego. This structure, so closely associated with our Morality, is a dissociative element which splits the affective basis of conscious and empathetic predominance in two, creating the homeostatic conditions for social control, neurosis, compulsive consumerism, cruelty, existential angst and unthinking obedience to authority. Standing in direct opposition to Ethical functioning which evolves as a natural product of identification based in the innervations of the sympathetic dopaminergic limbic/OFC circuitry, punitive morality finds its historical footing and epigenetic expression based in patriarchal penalty and mutilation (Norman 2013). The rebalancing of the circuitry involved is difficult, involved and painful (Norman 2011, 2013a). However, this single error has cast the unhappy lot of man and provided all those throughout history with a hopeless situation doomed to repeat itself. To examine the numerous wars in constant procession throughout recorded human history and understand that top down control of the affects is a clear sham and a consistent failure, is to understand that the basis of human empathy must be allowed its natural place as the progenitor of ethical behavior. Although the road for a modern adult is filled with pain to alter this pathological basis, the pathway for the next generations is a hopeful one (Norman 2015*f*). In raising the next generations in an environment free of reaction formations and penalties in excess, the basis of human connection and empathy may be nurtured and the native connectivity and kindness within man, might finally meet with his allotted measure of intellect. The broken race of man has within it the seed of its own ascension. The hope for mankind is an atavistic evolution.

References:

Bengston, M. (2015). Catatonic Schizophrenia. *Psych Central*. Retrieved on June 29, 2016, from: http://psychcentral.com/lib/catatonic-schizophrenia/

Bowlby, J. (1980) *Attachment and Loss. Volume 1, Attachment*. Basic Books, New York.

Brown, L. (2011) Intersubjective Processes and the Unconscious. Routledge, London.

Dodds, E. R. (1973). *The greeks and the irrational*. Los Angeles: University of California Press.

Freud, S. (1886-1939). *The standard edition of the complete psychological works of Sigmund Freud volumes one through twenty-four*. London: Hogarth Press, 2001.

Hopkins, J. (2013) Conflict Creates an Unconscious Id. *Neuropsychoanalysis*, 15, 45-48. http://dx.doi.org/10.1080/15294145.2013.10773718

Kaplan-Solms, K. and Solms, M. (2002) *Clinical Studies in Neuropsychoanalysis: Introduction to a Depth Neuropsychology.* Karnac Press, London.

Keveren, E.B., Martensz, N. and Tuite, B. (1989) Beta-Endorphin Concentrations in CSF of Monkeys Are Influenced by Grooming Relationships. *Psychoneuroendocrinology,* 14, 155-161.

http://dx.doi.org/10.1016/0306-4530(89)90065-6

Klein, M. (1952) *Some Theoretical Conclusions regarding the Emotional Life of the Infant.* In: The Writings of Melanie Klein, Volume 8: Envy and Gratitude and Other Works, Hogarth Press, London, 61-94. 487-501.
http://dx.doi.org/10.14704/nq.2015.13.4.869

Montagu, A. (1978) *Touching: The Human Significance of the Skin.* Harper and Row, New York.

Norman R. L. (2011) *The tangible self.* O'Brien, OR.: Standing Dead Publications.

Norman, R. L. (2013) Who Fired Prometheus? The Historical Genesis and Ontology of Super-ego and the Castration Complex: The Destructuralization and Repair of Modern Personality—An Essay in Five Parts. *The Journal of Unconscious Psychology and Self-Psychoanalysis.* www.thejournalofunconsciouspsychology.com

Norman, R. L. (2013*a*) Re-Polarization Theory: From Native Psychoanalysis to Sublimation—The Practical Reconstruction of Modern Personality. *The Journal of Unconscious Psychology and Self-Psychoanalysis*; File Retrieved
From: www.thejournalofunconsciouspsychology.com

Norman, R. L. (2013*b*) Nine Short Essays and Native Psychoanalysis—a Non-Elliptical Technique: Necessary Background Information Basic to Native Psychoanalysis. *The Journal of Unconscious Psychology and Self-Psychoanalysis*; File Retrieved
From: www.thejournalofunconsciouspsychology.com

Norman R. L. (2013*c*) Mind Body Syndrome—the unconscious constellation: Condensation, abreaction and dissociative-repression in the genesis and disbandment of Tension Myositis Syndrome. *The Journal of Unconscious Psychology and Self-Psychoanalysis*; File Retrieved From: www.thejournalofunconsciouspsychology.com

Norman, R. L. (2013*d*) Brahma and universal process identification: Enlightenment—a psychoanalytic perspective. *Mind* Magazine.

http://www.mindmagazine.net/#!new-ideas/czpl

http://www.mindmagazine.net/

Norman, R. L. (2014) Limbic Connectivity and Sympathetic Neural Balance: The Primary Psycho-physiological Locus of Affect. *Mind* Magazine.

http://www.mindmagazine.net/#!new-ideas/czpl

http://www.mindmagazine.net/

Norman, R. L. (2015) Quantum Unconscious Pre-Space: A Psychoanalytic Neuroscientific Analysis of the Cognitive Science of Elio Conte—The Hard Problem of Consciousness, New Approaches and Directions. *Neuroquantology*, 13, 4. doi: 10.14704/nq.2015.13.4.869

Norman, R. (2015*a*) (Semi)-Regressive Plastic Attachment Therapy. *Mind* Magazine. New Ideas section.

http://www.mindmagazine.net/#!new-ideas/czpl

www.mindmagazine.net

Norman, R. L. (2015*b*) Modern Man of Phylogeny, Guilt, Obedience and Consequence—An Answer to Old Problems. *Mind* Magazine. New Ideas section. http://www.mindmagazine.net/#!new-ideas/czpl

www.mindmagazine.net

Norman, R. L. (2015*c*) Mnemic Psycho-Epigenetics: The Foundational Basis of Depth, Archetype and Synthesis in Psychology. *Mind* Magazine. New Ideas section. www.mindmagazine.net

Norman, R. L. (2015*d*) The Epigenetic Unconscious pt. 1. *Mind* Magazine. New Ideas section.

http://www.mindmagazine.net/#!new-ideas/czpl

www.mindmagazine.net

Norman, R. L. (2015*e*) The Epigenetic Unconscious pt. 2. *Mind* Magazine. New Ideas section.

http://www.mindmagazine.net/#!new-ideas/czpl

www.mindmagazine.net

Norman R. L. (2015*f*) A new paradigm needs…a new myth. *Mind* Magazine. New Ideas section.

http://www.mindmagazine.net/#!new-ideas/czpl

www.mindmagazine.net

Norman, R. L. (2016) The Quantitative Unconscious: A Psychoanalytic Perturbation-Theoretic Approach to the Complexity of Neuronal Systems in the Neuroses, *Neuroquantology*, Vol. 14 issue 2 10.14704/nq.2016.14.2.949 **356-368**

Norman, R. L. (2016*a*) Homeostatic Conductance and Parasympathetic Basis Alteration: Two Alternative Approaches to Deep Brain Stimulation in Parkinson's, Obsessive Compulsive Disorder and Depression. *World Journal of Neuroscience*,

6, 52-61. http://dx.doi.org/10.4236/wjns.2016.61007

Norman R. L. (2016*b*) New therapeutic intervention and assessment tools: GSR, sexual dysfunction and the Peptide Assisted Therapy method—an applied therapy and mathematical metric of healing. *viXra*. http://vixra.org/pdf/1607.0117v1.pdf

Panksepp, J. (1998) *Affective Neuroscience: The Foundations of Human and Animal Emotions.* Oxford Press, New York.

Rizzolatti, G., Maddalena Fabbri-Destro, M. and Cattaneo, L. (2009) Mirror neurons and their clinical relevance *Nature Clinical Practice Neurology* **5**, 24-34 doi:10.1038/ncpneuro0990

http://www.nature.com/nrneurol/journal/v5/n1/full/ncpneuro0990.html

12. Nuclear Power and the World's Energy Requirements.

Introduction.

One of the more important factors involved when determining the overall development of a country is its energy consumption. It is undoubtedly the case that this factor provides a major difference between the so-called developed and under (or less) developed countries of the world. During the post-war period, the rapid development of the economies of the Western World was linked closely to oil, and possibly still is. Oil was used for a wide variety of purposes, for electricity production, for transport, as well as in the growth of the entire petrochemical industry. However, the oil crises of 1973 and 1979 produced a change in attitude and the main change was in the effort employed to make the West less vulnerable to the power of the major oil providers. This change did not affect the developed world too drastically, but the under-developed countries fared less well and many plunged even further into debt. With the population of the under-developed world being larger than, and increasing faster than, the population of the West, it seems the situation can only deteriorate.

In 1999, the United Nations announced that the world's population had reached six billion, a mere twelve years after reaching the five billion mark. It was predicted that the seven billion mark would be achieved between 2011 and 2015, with the actual outcome depending crucially on the situations in

China and India, the two most highly populated countries which between them account for a large percentage of the world's population. The reason these two countries are so important in any consideration of energy needs is because both are counted in the group of under-developed countries. This is of crucial importance because the so-called under-developed part of the world uses far less energy per head of population than does the developed part. Not too long ago, it was estimated that twelve times as much energy per person is used in the developed countries as compared with the under-developed ones. However, that situation has changed already and is set to continue doing so rapidly as this large group of countries strives to catch up with the rest. A further problem, which could increase in the future and must be of concern, is that much of this energy is provided by the combustion of fossil fuels[1], resulting in the production of large quantities of CO_2, SOx, and NOx, with the well documented attendant problems.

It was estimated[1] that, towards the beginning of this century, the energy consumption of the world was in the region of 2×10^{20} Joules per year, which equates to a rate of working of something of the order of 0.63×10^{13} Watts. With the world population being around six and a quarter billion at the same time, it followed that each person accounted for about 1kW. However, such a figure ignores the fact that most of the energy produced was used by those inhabiting the developed world – roughly 3.78 kW per person - while those in the under-developed parts of the world consumed approximately 0.315kW per head of population. It is reasonable, therefore, to expect the under-developed parts of the world to wish to achieve a more balanced state of affairs re energy consumption. Achieving this for them would improve so many aspects of life for so many people; in particular, health should improve and life expectancy increase. However, any successful modernisation would entail a huge increase in energy consumption, which would be exacerbated by any significant increase in the world's population. Hence, the problem of satisfying the world's energy needs was at the beginning of this century, and remains, a major one and still requires addressing urgently in an open-minded manner. This is so because the usual energy sources represent a finite energy reservoir and some of the thermodynamic implications of present practices need examining if a clean environment is to be produced for future generations.

Traditional sources of energy.

The reserves of fossil fuels are known to be finite and, even at the current level of usage, their life-times are fairly small. In fact, it might be noted that already in 1999 and the first quarter of 2002, the total world demand for oil exceeded the total world supply[2]. These two cases may be merely blips in the statistics but, nevertheless, sound a warning as far as dependence on oil is concerned. Coal, on the other hand, presents different problems. The stocks are diminishing rapidly, the cost of extraction in some cases is increasing and, like oil, it contributes considerably to the planet's environmental problems when used as a fuel. Another major source in the West is provided by natural gas which has the advantage of not producing high quantities of CO_2 when burnt, but its stocks are strictly limited. Furthermore, when the above population figures and the relative sizes of the developed and under-developed sections of the world are noted, it is seen that the energy requirements of the world are certain to rise drastically in the near future. This means that, even allowing for the possible discovery of new resources, fossil fuels will be unable to provide the world with sufficient energy for any significant length of time. It might be noted also that fossil fuels are used extensively in both the pharmaceutical and petrochemical industries, where substitutes prove expensive alternatives.

The unfortunately, and thermodynamically incorrectly, named 'renewable' energy sources, although quite numerous and varied, are unlikely to be able to contribute significantly more than about 20% of future total energy requirements[3]. These sources include geothermal energy, solar energy, wind power and wave power. Numerous though these may seem, it remains extremely unlikely that, taken together, these could combine to satisfy the world's future energy needs, especially if increased demand is accompanied by a decrease in the availability of fossil fuels as seems likely. All these sources of energy must surely have an important rôle to play, but it should always be remembered that while these sources are termed 'renewable', and although they truly seem non-decreasing, they too represent finite sources ultimately; -the second law of thermodynamics would allow nothing else!

It is well-known that, in the regions of the earth not too far from the surface, there is a temperature gradient of roughly 30K/km. In some places, the higher temperatures below the surface lead to geysers and other phenomena. However, the heat distribution is not uniform, with the temperature gradient being much greater in some places than others. A geyser is formed if water accumulates deep down where it is turned into steam which builds up in pressure before breaking through the earth's surface. Some of these naturally occurring phenomena have been harnessed to provide superheated steam which, in turn, may be used to provide power. Such plants may well make a useful contribution to energy needs but they are unlikely to make any worthwhile impact as far as global energy needs are concerned.

Wind and solar power, the two major regenerative sources, face the major problem of requiring a substantial portion of the earth's surface to provide the required energy. It has been speculated[1] that, at some time in the future, if the reliance on these two sources was increased, that portion could be 10% or more. What is more, such land surface would have to be in carefully chosen, appropriate places; possibly in the tropics for solar power or in known windy regions for wind power. There would also be associated transmission problems but, possibly more importantly, although wind and solar power sound attractive to many people initially, as soon as the amount of land to be committed to such schemes became known, it is likely that social objections would be raised quite forcibly. Further, both sources would be unable to guarantee actual production at any particular time and so substantial high power storage facilities would be needed and, as yet, no such facility exists. It has been estimated[1] that these two sources could not provide more than about 20% of Britain's energy requirements and possibly less for some other northern countries. These two sources must be remembered, however, as long term possibilities for at least helping provide for the world's energy needs. Further, as far as wind power is concerned, other queries have emerged since the enormous proliferation of wind turbines in Britain in recent years. The major one being to wonder why the design of turbine seen so often has been chosen. Such a turbine has a central axis to allow the whole to rotate to account for different wind directions. The actual turbine blades are then attached to another axis at right angles to the central one. Such a structure is obviously not totally stable as is evidenced by some graphic film posted on the internet. Also, such turbines only operate under a limited range of wind speeds – if not, at high speeds disaster could occur. This makes one wonder why the Finnish design of turbine which has only a central vertical axis and can operate under virtually all wind speeds was not chosen. Also, as alluded to earlier, since the supply of energy must be sporadic in nature, one must wonder how useful such an irregular supply would be to any national grid.

The harnessing of wave power presents its own set of seemingly enormous engineering problems and, so far, it seems there has been little progress in solving the practical problems of energy conversion associated with this form of power. However, looming over everything is the sheer power of the sea. It will be a truly tremendous feat of engineering to produce a device which is able to harness the power of the sea for our energy needs; a device that is robust enough to withstand major storm conditions and yet delicate enough to operate efficiently in conditions of relative calm. Any deployment of collectors for such a system would inevitably affect shipping and it is doubtful that any system would satisfy the worlds' total energy needs, at least not in the near future. However, this is certainly another potential source not to be forgotten.

Another source of energy, particularly popular in some parts of the world, is biomass. However, this source presents a big danger because its abuse could accelerate the world deforestation process. This source is another which should not be termed 'renewable' since, at present rates, for every ten hectares cut down, only one is being replanted. Another disadvantage with this fuel is that it provides another source of contamination of the atmosphere.

Other potential sources, such as ocean thermal power and the hydraulic resource, as well as further details of the above-mentioned sources have been discussed elsewhere[1]. It seems, unfortunately, that wave power, biomass, geothermal energy and tidal sources have all been found lacking when it comes to providing for the worlds' future probable energy needs; they provide insufficient power for present, leave alone future, purposes. This leaves the fossil fuels, which are slowly but surely disappearing, and nuclear power.

Nuclear Power.

At present, nuclear fission reactors provide a significant proportion of the world's energy. High concentrations of these plants are to be found in the U.S.A., Japan and Europe. However, once again there reliance on a finite source of fuel, uranium; although, in terms of power production potential, resources are much greater than is the case for fossil fuels. In many ways, as far as the projected time for which mankind might survive is concerned, one major sustainable method of energy production is provided by fast breeder reactors. In these reactors, under appropriate conditions, the neutrons given off by fission reactions can 'breed' more fuel from otherwise non-fissionable isotopes. The most commonly used reaction for this purpose is by obtaining plutonium 239Pu from non-fissionable uranium 238U. The term 'fast breeder' refers to the situations where more fissionable material is produced by the reactor itself. This latter situation is possible because uranium 238U is many times more abundant than fissionable uranium 235U and may be converted into plutonium 239Pu, which may be used as fuel, by the neutrons from a fission chain reaction. Attractive though such reactors may appear at first, they prove to be extremely expensive, largely due to important safety concerns surrounding the use of molten metals to remove the huge quantities of heat produced and to the fact that the fuel is highly radioactive plutonium. However, nuclear power always raises great worries with many people on at least two counts: firstly, there is always worry over a possible accident occurring, and secondly there is worry over the disposal of any radioactive waste. Countries such as the UK and Japan reprocess a proportion of the waste for use in weapons and medical facilities. However, this is expensive and time consuming and should be viewed as a form of recycling, rather than waste 'disposal'. In countries such as France and the USA, the majority of the waste is stored in water tanks on the actual sites of the nuclear fission reactors. This has led to a huge build-up, over the past fifty years, of a substantial stockpile of highly radioactive waste. This has prompted the need to find essentially permanent storage facilities for the material and, for example, the American government is having such a storage facility constructed at Yucca mountain in Nevada. This proposed facility is proving an enormously expensive exercise as reported in the National Geographic[4].

The big growth in the use of nuclear power came approximately thirty years ago and was probably due to the oil crises of the seventies. As soon as the price of oil returned to normality, however, nuclear energy ceased being competitive, mainly because of the high costs associated with basic nuclear technology. These costs are recoverable in the long term and proof of that claim is provided by realising that in 2002, the cost in cents per kWh of electric generation was 1.76 for nuclear power, 1.79 for coal, 5.28 for oil and 5.69 for gas; where these costs cover fuel, operation and maintenance, but not capital costs[5]. Hence, nuclear power is able to undercut other forms of energy generation and so should, in the longer term, be capable of recovering the initial capital outlay without losing the lowest position on the cost scale. It is always worth remembering also that, while there are drawbacks associated with the use of nuclear power, its use does not produce the dangerous gases which are polluting the atmosphere and causing acid rain. These may seem small points but everything needs to be taken into account when attempting to assess the provision of the world's future energy requirements.

Conventional methods for the disposal of radioactive waste.

Radioactive material that cannot be utilised directly in other processes is designated nuclear waste and most nuclear processes produce amounts of such waste. Long term solutions for its safe disposal have been sought for many years but, even today, few suggested solutions have been implemented. There are, in fact, several categories of waste but here attention will be restricted to a consideration of methods for disposing of so-called 'high level' waste. Modern conventional nuclear reactors (advanced gas reactors and pressurised water reactors) use enriched uranium fuel as a heat source. This is made from natural uranium ore which typically contains about 0.7% uranium ^{235}U, enriched to between two and three per cent, depending on the requirements of the particular reactor. This leaves a large amount of uranium ^{238}U with a reduced concentration of uranium ^{235}U; this is classed as 'medium level' radioactive waste. The enriched fuel is then compacted into fuel rods as $UO2$, ready for use in a reactor core. When exposed to 'thermalised' neutrons, the uranium ^{235}U

undergoes stimulated fission, leading to the production of a great variety of radioactive by-products, stored in the fuel rods. Once the concentration of uranium ^{235}U drops below about 0.9%, the fuel rod is classed as 'spent' and a new rod replaces it. The 'used' fuel rods produce a considerable amount of heat due to their high level of radioactivity -approximately 3×10^8 times that of a new fuel rod -and are stored typically in ten metre deep water pools on site for at least twelve months. This storage is to allow them to cool and for their radioactivity to decrease to a safer level. These 'cool' rods are then felt safer to transport and may be sent either to a reprocessing plant where useful products such as plutonium and the remaining uranium may be extracted or, more usually, may be moved to a large, longer-term storage facility.

The reprocessing of the fuel rods is achieved by cutting them up and dissolving them in nitric acid. This releases most of the gaseous fission products into solution; the exception being the noble gases. Most of the radioactivity in the spent fuel rods ($\approx 76\%$) originates from the fission products, except plutonium. Since the plutonium and remaining uranium are of use, they are removed from the solution chemically, leaving the highly radioactive waste in solution. This solution is then stored for a number of years before being evaporated and vitrified into glass blocks for long-term storage. This process, although seemingly efficient, in that the final waste material contains about 97% of the waste fission products, produces a large amount of low and intermediate waste which must be disposed of also. However, once waste is in the form of vitrified waste or cool fuel rods, it may be 'disposed' of either by being placed sufficiently out of harms way so that it requires no more monitoring or alternatively by being 'neutralised' by conversion to a harmless substance.

At present, the most popular method is to store the waste deep underground in very stable geological sites so that, by the time the waste leaks out, it is of no danger to life on earth. Such sites are required to be such that the waste may be safely stored for of the order of 400,000 years. One major problem with this, however, is that there is little evidence to support the supposition that the containers designed for the task would themselves survive for such a long time. There is also a great deal of controversy over the levels of seepage of radioactive elements from the stored waste, since predictions over such a long period of time are fraught with inherent uncertainties.

It is interesting, and possibly instructive, to consider data from what amounts to a natural uranium reactor, which provides a precedent for radio-isotope distribution over a very long time scale. A recently discovered site in West Africa had an unusually low concentration of uranium ^{235}U within the uranium ore. The only way it is felt this can be explained is if a significant proportion of the original uranium ^{235}U underwent fission. The area of land concerned is saturated with water which would provide a moderator capable of thermalising the neutrons. If the concentrations of uranium ^{235}U were sufficiently high, it is perfectly possible for a natural fission reactor to operate. Indeed, the concentrations of radioactive products indicate that this natural reactor operated approximately 1.8

billion years ago. When measurements were taken to see how far the metallic radioactive products had travelled in that time, it was found to be less than a metre from the original reactor site. Although the data is specific to the site in question, it does suggest that the level of transport of waste may be insignificant as far as the human race is concerned.

Another method of dealing with radioactive waste, which is under consideration at present, is the conversion of the waste into less dangerous materials, usually through high intensity neutron bombardment. The idea is currently still at the development stage but its main disadvantage is the low volume of waste that can be practically converted in this way.

An alternative method for disposal of high-level radioactive waste.

An alternative method for disposing of high-level radioactive waste has been proposed recently by Santilli[6]. It is a form of neutralisation but does not use the conventional methods currently being researched. Indeed, classical formulations of quantum chemistry and nuclear models do not even permit the practical method proposed. This new method arises from a number of discrepancies between the theoretical and measured values using the current formulation of quantum mechanics. Santilli has attempted to resolve these issues by formulating what might be termed a new form of quantum mechanics, known as hadronic mechanics, which is based on a new type of mathematics called isomathematics[6]. Although abstract in nature, isomathematics has already had some definite practical success. For example, it has been used successfully to predict the growth of sea shells, something which could not be done previously using conventional mathematical techniques[7]. Though only mentioned in passing, hadronic mathematics is an extensive rewrite of theory as known by most people. It is not, however, excessively complex, merely different and it is that that initially makes it hard to grasp. However, once the basic formalism is understood, much of what may be deduced follows quite straightforwardly. If this new theory is a true representation of nuclear and molecular structure, then it predicts that neutrons may be viewed as compressed hydrogen atoms. Conventionally, the probability for beta-decay of a neutron into a proton, electron and neutrino is very low for radioactive elements on a nuclear timescale; for stable isotopes, the lifetime of neutrons is effectively infinite. Hadronic mechanics predicts that such a reaction may be stimulated within the nuclei of radioactive materials.

In essence, a radioactive nucleus is in an excited energy state and is attempting to return to its ground state energy. Under normal circumstances, this is achieved by spontaneous fission or radioactive emission, the time taken to decay being dependent on how much excess energy the nucleus has. This can vary between 10^{-31} seconds and millions of years. An excited nucleus can return to its ground state through emission of a photon (gamma emission), an electron (beta emission), or by spontaneous fission, where alpha emission is assumed to be a form of fission. The latter two processes cause a change in the nature of the parent nucleus, altering its nuclear properties. The energy value of the excited state determines the method by which the nucleus returns to its ground state. If the decay process involves the emission of a beta particle, it may be extrapolated that a neutron will have to decay to achieve this.

From the theoretical calculations, it is hypothesised that this decay can be stimulated by bombarding the nucleus with so-called 'resonant' photons with an energy of 1.294 Mev. Under normal circumstances the probability of this interaction is extremely low. However, Santilli claims that there is a large resonance peak in the reaction cross-section (that is, the probability of the said interaction occurring) for incident photons with an energy of 1.294 Mev. It is also feasible, though not stated, that the simple existence of an excited nucleus makes it open to interaction with resonant photons, regardless of the means of decay ultimately used to return to its ground state energy. Once a neutron is converted into a proton plus reaction products, a number of possibilities could occur. Firstly, the new nucleus could be a stable isotope, in which case further interactions with the resonant photons would be unlikely and the waste would have been effectively neutralised. Secondly, the new isotope could form a new neutron deficient nucleus and one of the following could then occur: the nucleus undergoes spontaneous fission, forming two new nuclei and possibly a number of neutrons, which could interact with other fissile elements in the fuel and generate excess heat; the neutron deficient nucleus could form a new excited energy state which can simply be categorised as another target radioactive nucleus for the resonant photons.

If this interaction is found to be true, its application for the disposal of radioactive waste is profound. Photons with the correct resonance energy can be produced easily within a piece of equipment of small volume, such that the neutraliser could be built on the same site as the parent reactor itself. Effectively, it would allow all radioactive waste to be fissioned until all the isotopes form stable nuclei. However, a point to note is that, taking a typical sample of waste, the resultant treated material would not be radioactively dangerous but chemically could be a totally unknown concoction of elements and compounds, which may well contain high levels of toxins. Another point to note is that stimulated fission would release a considerable amount of heat energy from the fuel, and so some sort of effective coolant would be required. However, since this heat energy could be used to produce even more power, there seems no reason in principle to suppose that what might be termed a secondary 'waste reactor' could not be built.

To continue quantitative scientific studies of the proposed new method for the disposal of nuclear waste essentially requires three basic experiments to be performed. All should be of reasonable cost and are certainly realisable with present-day technology. Firstly, the experiments of Rauch and his associates[8], in which direct measurements of the alterability of the intrinsic magnetic moments of nucleons were made, should be repeated and to as high a degree of accuracy as possible. Secondly, don Borghi's experiment[9] on the apparent synthesis of the neutron from protons and electrons only should be repeated also. It is interesting to realise that, despite enormous advances in knowledge in recent times, fundamental experimental knowledge of the structure of the neutron is missing still. Finally, it is necessary to determine whether or not gamma stimulated neutron decay will occur at the resonating gamma frequency of 1.294Mev. One way of achieving this is to have Tsagas's experiment on stimulated neutron decay[10] completed. However sceptical someone may be of these new ideas, it seems sensible to perform these experiments to decide if they are valid or not. If they are valid, the rewards would be tremendous; if not, little would have been lost.

Even assuming that the theory is found to be sound and the predicted resonance peak exists, there would still be further practical considerations when applied to the disposal of radioactive waste. Nevertheless, it is easy to see that, if proven, such a method would save a truly considerable amount of public funds, given the relatively low cost of the apparatus as compared with the removal of the need to transport the spent fuel to reprocessing facilities and also with the building of long-term storage facilities. The possibility of producing toxic by-products is, however, a real concern and a means for the disposal of such by-products, if they did materialise, would have to be sought as a matter of urgency.

Conclusion.

Hence, the world faces an almost exponentially increasing demand for energy due to the underdeveloped sections of the world becoming more industrialised and demanding an improved standard of living and this position is exacerbated by the rapid increase in the worlds' population. This ignores the possibility of a further increase in demand due to the introduction of new technology. This demand cannot be met by the use of fossil fuels and, in any case, if it could, the increased use of such fuels would surely have a less than beneficial effect on the environment. The regenerative and so-called 'renewable' forms of energy production are seen to be able to make a contribution, particularly locally, but they do not seem capable of having a truly major effect. Although not mentioned previously, it may be noted that the constructing of a first nuclear fusion reactor seems as far away as ever; indeed, many feel such a reactor impossible to build. It seems, therefore, that, with the existing state of human knowledge, the only viable energy source sufficient for supplying the future energy needs of the world is nuclear power. It has to be recognised that there are attendant problems. People are, and probably will be for a long time, very uneasy about nuclear power. They've seen its awful potential destructive power and so, quite naturally, worry about the possibility of accidents, even catastrophic accidents, at the plants themselves. People are also very well aware of the major problem posed by nuclear waste. Although the traditional methods of dealing with this waste are acceptable, they are politically controversial and/or extremely expensive in monetary terms, both factors being highly important in the case of the location of underground storage facilities. Various others methods have been advocated over the years but not one has remained in favour for long. Here attention has been drawn to the relatively new ideas proposed by Santilli. They are revolutionary in concept, they do draw on a new form of mathematics and quantum mechanics but tests have been carried out already to see if the theory works. More tests are being carried out but the initial results are positive. If the ideas are eventually proven, they will provide the possibility for a means of radioactive waste disposal which satisfies the requirements for convenience, finality of disposal, political acceptance and cost. As with all new ideas there is scepticism within the existing scientific community but, if Santilli's theories are finally supported by experimental evidence, few grounds for objection could remain for what could be a revolutionary technology. It is to be hoped that experimentation to validate, or otherwise continue Santilli's theories will be performed.

References.

[1] G. H. A. Cole; 1996, in *Entropy and Entropy Generation*, J. S. Shiner (ed), pp 159-173, Kluwer Academic Publishers, Dordrecht.

[2] www.eia.doe.gov/emeo/ipsr/t21.xls

[3] M. Scott & D. Johnson; 1993, *Nuclear Power*, Open University.

[4] *National Geographic,* July 2002.

[5] *Access to Energy,* September, 2002.

[6] R. M. Santilli; 2001, *Foundations of Hadronic Chemistry* and references cited there, Kluwer Academic Publishers, Dordrecht.

[7] R. M. Santilli, 1996, *Isotopic, Genotopic and Hyperstructural Methods in Theoretical Biology*, Naukova Dumka Publisher, Ukraine.

[8] H. Rauch et al; 1975, Phys. Lett. **A54,** 425.

[9] C. Borghi; 1993, J. Nuclear Phys. **56**, 147 (in Russian)

[10] N. F. Tsagas; 1996, Hadronic J. **19**, 87.

13. Some Comments on Possible Causes of Climate Change.

Introduction.

Climate change is quite possibly the most hotly debated issue at the moment and there are many conflicting views about the causes. As well as being an issue that affects most of us on a daily basis, it is a very important political issue.

So what is happening to convince people that there is a problem? The BBC news reports that average global temperatures have risen by 0.7° C over the last 300 years. 0.5° C of that warming occurred during the 20th century, and most of that occurred between 1910-1940 and from 1976 onwards. Four out of five of the warmest years ever to be recorded were in the 1990's, with 1998 being the warmest year globally since records began in 1861. It is also widely believed that arctic sea ice is thinning and that there has been an average increase of between 0.1 and 0.2 meters in sea levels globally over the last 100 years. In many high and mid-level areas in the northern hemisphere, precipitation has increased by 0.5-1% per decade. Finally in Asia and Africa the frequency and intensity of droughts has increased in the last few decades. (BBC08)

The Intergovernmental Panel on Climate Change (IPCC) begins its *Climate Change 2007: Synthesis Report* with the statement "Warming of the climate system is unequivocal, as is now evident from the observations of increases in global average air and ocean temperatures, widespread melting of snow and ice, and rising global average sea level". According to the IPCC: temperature increase is widespread over the globe and greater at higher northern latitudes, with land regions warming faster than the oceans. Global average sea level has risen since 1961 at an average rate of 1.8 [1.3 to 2.3][1] mm/year and since 1993 at an average rate of 3.1 [2.4 to 3.8] mm/year, due to thermal expansion, melting glaciers and icecaps, and the polar ice sheets. It follows that observed decreases in snow and ice extent are also consistent with warming. Satellite data has shown that arctic sea ice has shrunk by 2.7 [2.1 to 3.3] % per decade with larger decreases in summer of 7.4 [5.0 to 9.8] % per decade. Also mountain glaciers and snow cover on average have declined in both hemispheres. They state that there is a *very high confidence* that the earlier timings of spring events and poleward and upward shifts in plant and animal ranges are linked to recent warming. In some marine and freshwater systems, shifts in ranges and changes in algal, plankton and fish abundance are with *high confidence* linked to rising water temperatures as well as ice cover, salinity, oxygen levels and circulation. Of the 29,000 plus observational data series, taken from 75 studies, which show significant change in physical and biological systems, more than 89% are consistent with the change expected as a response to warming. Nevertheless, there is a geological imbalance with a notable lack of data and literature on changes coming from developing countries. (IPCC, 2007)

However, according to reports from the U.S. National Oceanic and Atmospheric Administration (NOAA), almost all the allegedly 'lost' ice is back, they show that ice which had shrunk from 13 million square kilometres in January 2007 to just 4 million in October is almost back to its original level and figures show that there is nearly a third more ice in Antarctica than is usual for this time of year. Scientists are saying that last winter, the northern hemisphere endured its coldest winter in decades and that snow cover across that area was at its greatest since 1966. One exception to this was Western Europe which experienced unseasonable warm weather; the UK reported one of its warmest winters on record. Vast parts of the world have, however, suffered chaos because of some of the heaviest snowfalls in decades, including central and southern China, the United States and Canada. In China, the snowfall was so severe that over 100,000 houses collapsed under the weight of it. Jerusalem, Damascus, Amman and Saudi Arabia all reported snow and below zero temperature and in Afghanistan snow and freezing weather killed 120 people. (Bonnici, 18.02.08) (Brennan, 19.02.08)

It is clear that our planet's climate is changing and some opinion suggests that this is down to us. But some scientists believe that because the climate has changed naturally before that it is supposed to change and therefore disagree that there is even a problem. Our climate is an incredibly complex system and there is doubt over whether enough is known about it to make predictions and whether the computer models that are being used are adequate. (BBC08)

Possible causes.

The atmosphere that surrounds the Earth plays an essential role in making our planet habitable, it is transparent to the visible radiation emitted by the Sun and this heats the Earth's surface, without it the temperature would soar by day and plummet by night, and the average temperature would be around -18°C. About 30% of the Sunlight that reaches our planet is reflected back to space by clouds, dust or the ground, more than 20% is absorbed in the atmosphere and almost 50% is absorbed by the Earth's surface. Some of the infrared radiation that is radiated by the Earth's Sun-warmed surface escapes through the atmosphere directly into space but most of it is absorbed on the way by clouds and greenhouse gases which release part of that heat into space and radiate some back to the surface increasing the temperature in the lower atmosphere. As the temperature of the Earth's surface rises, the amount of IR radiation increases. The temperature adjusts until a delicate balance is achieved. Unlike the two main components of air, oxygen (20%) and nitrogen (78%) that have a linear diatomic structure, greenhouse gases in the atmosphere such as water, carbon dioxide and methane have three or more atoms which make them well suited to absorbing radiation. As these greenhouse gases accumulate they block each other's radiation to space and so, the more greenhouse gas there is, the warmer it gets. The average height at which the radiation can now escape to space then begins to increase and at higher altitudes the temperatures are cooler and radiation into space decreases. The system then starts to readjust, more water evaporates from oceans and lakes and sea ice which reflects Sunlight back into space begins to melt, reducing the reflection, these both amplify the warming effect. (Henson, 2006)

Another theory for the cause of global warming, developed by Henrik Svensmark amongst others, is that the Sun and cosmic rays play a role in the change in our climate. They believe that cosmic rays are an essential ingredient, which experts have so far been slow to appreciate. The cosmic rays must break through three defensive shields before they can reach the Earth's surface, first the Sun's magnetic field then the Earth's magnetic field and finally the air around us and only the most energetically charged particles can get as far as sea level. In Svensmark's theory it is these energetically charged particles called muons or heavy electrons, which are produced when cosmic rays hit the atmosphere, that help clouds to form low in the air and cool the Earth. Whist some clouds higher up can have a warming effect; these clouds which are less than 3000 meters high keep the Earth cool. Put simply this means more cosmic rays, more clouds and cooler temperatures. (Svensmark, et al., 2007)

The clouds play a very important role in our climate, about 60% of the globe is covered by cloud and we all appreciate how important cloudiness is in determining the temperature on a day to day basis. Clouds modulate the Earth's radiation balance, both in the visible and infrared spectra. Clouds cool the Earth by reflecting incoming Sunlight, the tiny drops in clouds can scatter between 20% and 90% of the Sunlight that reaches them, a cloud free earth would absorb nearly 20% more heat from the Sun than it does at present. However, the clouds also have a warming effect on the Earth, they absorb the infrared radiation emitted from the surface and reradiate it back down. This process traps heat like a blanket and slows the rate at which the surface cools down. The clouds reflect about 50Wm^{-2} of solar radiation up to space and radiate around 30Wm^{-2} down to the ground, the net effect being 20Wm^{-2} cooling on average. This greatly exceeds the 4Wm^{-2} warming due to atmospheric carbon dioxide levels doubling from 300 to 600ppm. What we don't know, however, is what the net cooling or warming effect of all clouds on Earth will be in changing atmosphere or how the clouds themselves will be changed by a change in the temperature of the Earth. If the cooling effect of clouds increases more than the heating effect does, the clouds would reduce the magnitude of the greenhouse-induced warming but speed its arrival. This is called negative feedback. Both effects decreasing could have the same effect but, if the cooling decreases more than the heating, the cloud changes would boost the magnitude of the warming but delay its arrival. In any scenario, the important factor is the net effect of the clouds. To complicate matters, however, the altitude of the clouds has an influence. High clouds have a net warming effect, they block little incoming solar radiation but, because they are at low temperature, they return little outgoing infrared radiation to the Earth's surface. Clouds at low altitude have a net cooling effect because they have a high albedo and being at a temperature which is nearly as warm as the surface of the Earth they emit nearly as much infrared radiation to space as the surface would under clear skies. (Rossow, et al., January, 1995)

There is also well established evidence that the Three Milankovitch Cycles in the Earth's rotation and orbit, change the amount and alter the distribution of Sunlight over the Earth, heating and cooling the Earth over cycles of 100,000-41,000 and 23,000 years. (Page, 27.06.07) The Milankovitch cycles is the name given to the collective effects of changes in the Earth's movements on the climate. The eccentricity, axial tilt and precession of the Earth's orbit vary in several patterns which have resulted in 100,000 year ice age cycles over the last few million years. The Earth's tilt goes up and down ranging from about 21.8° to 24.4° and back over an approximately 41,000 year cycle. The tilt is currently around 23.4° and decreasing. When the tilt is most pronounced it gives rise to stronger summer Sun and weaker winter Sun. Ice ages often occur because as the tilt decreases the progressively cooler summers cannot melt the past winter's snow. The Earth's orbit around the Sun is not precisely circular but elliptical in shape, with the Sun positioned slightly to one side of the centre point. The eccentricity, or 'offcenteredness' of the orbit varies over time in a complicated way, the result is two main cycles, one averages about 100,000 years in length and the other 400,000 years. When the eccentricity is low there is little change through the year in the distance between the Earth and Sun. When the eccentricity is high the Sunlight reaching the Earth can be more than 20% stronger at perihelion (currently January) than at aphelion (currently July). The Earth's axis also plays a role, the main cycle of the rotation around the axis is known as the precession, and takes about 26,000 years. It shifts the dates of the perihelion and the aphelion forward by about one day every 70 years. Currently the Earth is 3% closer to the Sun in early January (perihelion) than in early July (aphelion) with about 7% more solar energy reaching the Earth at perihelion. In about 13,000 years the Earth will be closest to the Sun in July instead of January; this will intensify the seasonal changes in solar energy across the Northern Hemisphere and weaken them in the South.

CO_2 emissions.

It is reported that man is producing and releasing into the atmosphere higher levels of carbon dioxide through increased industry and de-forestation and that this is affecting our climate by adding to the layer of greenhouse gases in the atmosphere. However, there are many factors which affect the levels of greenhouse gases in the atmosphere; the sea for example absorbs carbon dioxide.

Geologist Dr Norman J Page published an article in June 2007 entitled 'Climate Change and Carbon Dioxide' for the Alpha Institute for Advanced Study saying "As a geologist, I find the current climate of fear in which the debate on Global Warming is conducted very alarming". He starts by pointing out some common misconceptions often reported in the media:

The United States is often referred to as the world's biggest polluter but, whilst the U.S. does emit a large amount of CO_2, the land use patterns means that they also absorb a large amount and it is the net amount not the amount emitted which is the important figure. It has been shown by a paper published in Science Magazine in 1998 that, when the net amount of carbon dioxide is taken into account, North America actually takes up more carbon dioxide than it emits by about 100 million tonnes per year, whilst Europe emits a large amount overall. Carbon dioxide is often reported as being the biggest offender but in fact water vapour is Earth's most abundant greenhouse gas and CO_2 makes up less than 3%. Dr Page states that the Earth is now impoverished in CO_2 and that at various times in the last 550 million years levels of CO_2 have often been four or five times the current levels and at times ten to fifteen times greater.

Annually, human contribution to greenhouse gas is said to be between one and two tenths of a percent and it has been suggested that termites alone produce ten times more greenhouse gas than humans. It has also been reported that the amount of CO_2 produced by the population of India breathing is more than all of the coal burning plants in the U.S. Page states "If we eliminated human use of fossil fuels entirely it would have little impact on future temperatures". Dr Roth, at MIT, has shown that over time scales of more than 10 million years it is very difficult to prove a connection between the climate and CO_2. There is, however, a connection between temperature and CO_2 over shorter time intervals but data extracted from ice cores show that a natural warming period precedes a CO_2 increase and is not caused by it. (Page, 27.06.07)

Recent data from Britain's Climate Research Unit shows that the global mean temperature in October 2007 was 0.159 degrees lower than in October 1997; there has been no warming in the last 10 years but, in this same period, the levels of CO_2 have increased by 6%. Further, recent data from the Met Office also shows that the average global temperature for the first quarter of 2008 was cooler than the average of any year since 1996, whilst levels of CO_2 have risen by 6% (Page, 14.04.08). On the other hand the IPCC states that eleven of the last twelve years, from 1995 to 2006, rank among the twelve warmest years in the instrumental record of global surface temperature since 1850.

In November 2007, the Parliamentary Office of Science and Technology periodical Postnote published the report *Climate Change Science*. It stated that the atmospheric concentration of CO_2 has increased from a pre-industrial concentration of about 280 parts per million (ppm) to 379 ppm in 2005 and that over the last 650,000 years CO_2 varied within a range of 180 to 300 ppm, but there were approximately 90,000 measurements of CO_2 levels made since 1812, through the 19th and early 20th century which show that levels of CO_2 were at about 440 ppm in the 1820s and 1940s and about 370 ppm in the 1850s. The historical chemical data shows a clear trend, with the changes in CO_2 tracking the changes in temperature and therefore the climate. These measurements were made by chemists, several of whom had Nobel Prize level distinction, and the chemical methods used to make the measurements are good enough to measure with high accuracy, for example the Pettenkofer process was the standard analytical method for determining atmospheric carbon dioxide levels between 1857 and 1958 and usually achieved an accuracy better than 3%. This data was recently published by Ernest-Georg Beck, but modern climatologists have generally ignored the historic figures, discrediting the methods as unreliable despite the techniques being standard textbook procedures in several different disciplines. Furthermore this data was recently not acknowledged by the IPCC when it published its findings. (Beck, March 2007)

The factors that drive climate change are separated into two categories, forcings and feedbacks. Changes in solar energy output or in the concentration of greenhouse gases, natural or otherwise, are classed as forcings. Scientists quantify and compare the contributions of different agents that affect surface temperatures by measuring their 'radiative forcing' which can be positive or negative and is measured in Watts per metre squared. Feedbacks are internal climate processes that amplify or reduce the climate's response depending on how responsive the climate is to various forcing processes, for example, a warmer atmosphere can hold more moisture which itself acts as a greenhouse gas causing further warming, this is a positive feedback.

Postnote reports that scientists have estimated the combined human-caused radioactive forcing to be $+1.6Wm^{-2}$, and have decided that it is extremely likely that humans have exerted a substantial warming influence on the climate, and that this radioactive forcing is likely to be at least five times greater than the radioactive forcing due to solar changes. Solar irradiance is estimated to have caused a small warming effect $+0.12Wm^{-2}$. It concludes that, for the period from 1950 to 2005, it is extremely unlikely that the combined natural radioactive forcing (solar plus volcanic sources of aerosols) has had a warming influence that is comparable with that of the combined human made radioactive forcing. But there is an uncertainty in the predictions of the future climate state that is due to the uncertainty in the magnitude of climate forcings and feedbacks. The IPCC has classed the level of scientific understanding (LOSU) as low for the climate forcings of Solar irradiance, contrails from aircraft, cloud reflectivity and water vapour from the methane in the upper atmosphere.

Scientists use climate models to help quantify these forcings and feedbacks to predict the future state of the climate. Mathematical equations of the climate are fed into a three dimensional grid of points that cover the Earth's atmosphere and oceans. The resolution of the model, which is essentially the spacing between the points on the grid, is limited by the power of the computer used by the researcher, but is generally less than 150km in the horizontal direction and 1km in the vertical, with finer resolution near the surface of the Earth. Some scientists are confident that climate models provide credible estimates of future climate change at least on large scales, because their design is based on established physical laws which are used in weather forecasting models. The models have also been used to reproduce features of past climates and climate changes. But these models do have their limitations, some processes such as cloud formation, occur on time and space scales which are too small for the climate models to resolve. Modellers deal with this problem by representing small scale processes with average values over one grid box, but these assumptions are often a big source of climate model uncertainty. Climate modellers run a number of programs and average the results, hoping to try and remove or reduce some of the effects of natural variability, leaving the human caused changes. (Smith, November 2007) (Page, 29.11.07) (Page, 14.04.08)

The Intergovernmental Panel on Climate Change is a United Nations organisation and is there to help policymakers decide how to respond to climate change. Its role is not to carry out any scientific investigation but to evaluate studies carried out by thousands of researchers around the world and then synthesise the results into one report. For the 2001 and 2007 reports there were three working groups to deal with: the basis in physical science, impacts, adaptation and vulnerability and mitigation, each group is headed by a pair of scientists, one from a developed and one from a developing country. Climate Change 2007: Synthesis Report is based on the assessment carried out by the three working groups. The report starts by discussing the observed changes in the climate and their effects, they say it is *very likely* that over the past 50 years hot days and hot nights have become more frequent over most land areas and cold days, cold nights and frost have become less frequent. From observational evidence from all continents and most oceans it seems that many natural systems are being affected by regional climate changes, particularly temperature increases.

The report then goes on to discuss the causes of the change. The energy balance of the climate system is altered by changes in the atmospheric concentrations of greenhouse gases and aerosols, land cover and solar radiation. Levels of global greenhouse emissions due to human activity have grown since pre-industrial times and there was an increase of 70% between 1970 and 2004.

The atmospheric concentration of CO_2, 379 ppm, far exceeds the natural range over the last 650,000 years and this increase is attributed primarily to fossil fuel use. At 1774 ppb the atmospheric concentration of methane also exceeds the natural range over the same time period and it is very likely that this is predominantly due to agriculture and fossil fuel use. The IPCC concludes that there is a very high confidence that the net effect of human activities since 1750 has been one of warming and that continued greenhouse gas emissions at the current rate, or above, would cause further warming and induce changes in the global climate system that would very likely be larger than those observed already.

It is often reported that there is a consensus of opinion amongst the leading climate scientists that greenhouse gases are the cause of climate change but this is not so. According to many involved with the IPCC, including Professor John Christy, lead author, not all of the 2500 plus top scientists listed as contributors agree with the findings of the report, and some have had to fight to have their names removed. There have even been claims that the IPCC has censored its scientists. Professor Frederick Seitz wrote a letter to the Wall Street Journal saying that the version of the IPCC's latest report that was released was not the version approved by the scientists listed on the title page. He claims that at least 15 of the key sections in the science chapter have been deleted including the following statements:

"None of the studies cited above has shown any clear evidence that we can attribute the observed [climate] changes to the specific cause of increases in greenhouse gases"

and

"No study to date has positively attributed all or part [of the climate change observed to date] to anthropogenic causes".

The IPCC did not deny removing any sections but said there was "no bias" in their report. (Durkin, 2007)

American politician Al Gore is a passionate believer that greenhouse gases are the cause of climate change and has been campaigning world-wide for change. The documentary *An Inconvenient Truth* comprises the lectures he has been giving around the world on climate change as well as interviews with Gore. He claims that there is no doubt that the CO_2 emitted by humans thickens the layer of greenhouse gases and that this warms the planet. However, the documentary focuses more on the effects that global warming is having on the climate and the weather than showing that it is caused by CO_2. Gore says that the idea that the world's temperature has been warmer than it is at present is untrue and shows a graph which depicts the temperature during the Medieval Warming Period as being much cooler than that of today. He discusses the ice core method used to measure levels of temperature and CO_2 at periods in the past but does not mention what conclusions have been drawn from the results obtained. Gore then uses a graph showing the levels of CO_2, going back millions of years, which shows that levels have never exceeded 300 ppm until now but which does not agree with the data published by Beck. According to *An Inconvenient Truth*, the poles experience a much greater impact due to global warming claiming that, because earth's climate is non-linear, a global rise of 5°F would cause a 1°F rise at the equator but a 12°F rise at the poles. During the '*Inconvenient Truth*' lecture, Gore used images of icebergs breaking off and falling into the sea, ice sheets melting and a polar bear clinging to a melting iceberg; all images which have been used widely to show the need to fight climate change. It is claimed that our global warming is melting icebergs which is causing polar bears to drown in the oceans. However, he fails to mention that the polar bear population has soared in recent years or even that many of the photographs were taken in August when melting is normal. Gore's solution to the problem of global warming is to reduce carbon emissions by such measures as using so-called renewable energy and increasing both electricity and transport efficiency. Whilst the message of the documentary is conveyed strongly and emotively, using footage of natural disasters and stories of his own personal tragedies, the scientific content seems questionable, many of the graphs shown do not have axes or scales and much of the data quoted has no units. (Gore, 2007; Brennan, 19.02.08)

In response to *an inconvenient truth,* Martin Durkin made the documentary The Great Global Warming Swindle, which was shown last year on Channel Four. It features interviews and opinions of many top climate scientists who disagree with the greenhouse gas explanation of climate change. The documentary begins by discussing a few of the points in history where the temperature was significantly higher than that of today, the Medieval Warm Period, before the Little Ice Age for example and the Holocene Climatic Maximum, when the temperature was significantly higher than it is now for three millennia. The warming event seems to have peaked between 11,000 and 8,000 years ago and North-western Europe experienced warming, while there was cooling in the south. There is evidence at 120 of 140 sites across the western Arctic of temperatures warmer than at present. At 16 sites where quantitative estimates were obtained they show local temperatures that were on average 1.6±0.8 °C higher than present. The cause of this event is believed to be the Milankovitch cycles and a continuation of changes that caused the end of the last glacial period. When the axial tilt was at 24° and the Earth at its nearest approach to the Sun the warming would have been at its maximum in the Northern Hemisphere. The Milankovitch forcings would have provided 8% more solar radiation, calculated to be +40W/m² to the Northern Hemisphere in the summer, causing greater heating. (Durkin, 2007) (Davis, et al., 2003) (Kaufman, et al., 2004)

The data the IPCC used to demonstrate climate change, as well as that used by Al Gore and many others, indicates a period of intense warming during the early 20th century. Clearly in the period between 1905 and 1940 industry was fairly primitive. However, in the years after the 2nd World War when the world economy boomed, industry thrived and CO_2 levels soared but according to the data, from approximately 1940 to 1960 the world cooled. So it seems that the time does not fit with greenhouse warming.

It is known that incoming radiation from the Sun is trapped by the greenhouse gases in the troposphere, so it follows that, if an increasing level of greenhouse gases is responsible for the warming, the rate of warming should increase the higher you go. The temperature of the atmosphere may be measured using a satellite or by weather balloon. Using both methods it has been found that the rate of warming is in fact higher at the surface than the upper atmosphere, which does not fit the theory.

It is often reported in the media that a temperature rise of just a few degrees could have a huge warming impact, melting the icecaps. But records show that Greenland has been much warmer than it is today and did not have a big warming event.

Pictures of pieces of ice breaking off and falling into the sea are also often shown, but the icecaps are always naturally expanding and contracting and it is perfectly normal for pieces to break off.

Dr Ian Clarke used ice drilling to try and find out if there is a link between CO_2 and the climate and although he found a connection it was an unexpected one. The temperature on Earth leads the levels of CO_2 by 800 years. Several major ice surveys since have confirmed these findings. This can be explained by looking at the oceans. They contain CO_2 and each year they absorb some and emit some, but how much depends on the temperature. When the temperature is warm they release more and when it is cooler they absorb more. The oceans are so big that it takes a long time for them to warm or cool, often hundreds of years, which explains the time lag between the temperature and level of CO_2.

There is also doubt raised in the documentary about the accuracy of the climate models used. A model is only as good as the assumptions on which it is based. These climate models are based on hundreds of assumptions and it only takes one to seriously distort a models findings. The other concern with the current climate models is that they assume CO_2 is the main climate driver and do not incorporate any other possible influences. It is claimed that tweaking the parameters of the model, even slightly, can show a number of possible outcomes.

CO_2 makes up 0.054% of all the gases in the atmosphere, the percentage of anthropogenic CO_2 is even smaller and it is known that CO_2 accounts for just 3% of greenhouse gases which are themselves only a small part of the Earth's whole climate system. So what do the greenhouse sceptics believe is driving the changing climate? The Sun!

Sun spots are intense magnetic fields which appear at times of increased solar activity but even before this was understood astronomers would count the number of Sun spots in the belief that more heralded warmer weather. Edward Maunder noticed in 1893, during the Little Ice Age that there were barely any Sun spots visible, this became known as the Maunder Minimum. Eigil Friis-Christensen compared the Sun spots with temperature over the last 120 years and found a very close correlation, using astronomical data for the past 400 years the comparison was taken further back and was found to be intimately linked.

Astrophysicists from Harvard University conducted a study, in 2005, into the temperature-carbon dioxide and temperature-Sun relationships. No obvious link was found linking temperature and carbon dioxide, although a close link on a decade to decade basis was found between temperature and the Sun; this was based on independent data from NASA and the U.S Oceanic and Atmospheric Administration, displays. (Durkin, 2007)

Cosmic Rays.

In February 2007 A&G magazine-News and reviews in Astronomy & Geophysics published the article *Cosmoclimatology: a new theory emerges* by

Henrik Svensmark. This report contains a lot of in depth information on the research finding by Svensmark and emerging scientific evidence. The Sun's influence on our climate had long been recognised, by Herschel in 1801, Eddy in 1976, Friis Christensen and Lassen in 1991, but Henrik Svensmark and Eigil Friis Christensen noted the link between this relationship and the small 0.1% variations in the solar irradiance over a solar cycle measured by satellites. The pair published a paper entitled "Variation of cosmic ray flux and global cloud coverage-a missing link in solar-climate relationships" in 1997 after they announced their findings at the COSPAR space science meeting in Birmingham in 1996. (Svensmark, February 2007)

The Chilling Stars-A new theory of climate change was published in 2007 by Henrik Svensmark and Nigel Calder. The theory that cosmic rays play a role in the change in our climate has recently been developed and this book goes into detail to explain the theory and the evidence for it. It stems from Svensmark's research at the Danish National Space Centre. The book is a collaboration between Svensmark, who contributed the majority of the scientific input, and Calder who wrote it up. The pair met in 1996 when they were introduced by Friis-Christensen, who has also contributed toward the cosmic ray theory. They continued to discuss the theory over the following years and decided to collaborate on a book. The text evolved in 2005/06 whilst the intensive research continued, despite difficulties with funding.

Our Ancestors sometimes believed that the Moon and stars sucked heat from the Earth, but astronomers now know that most of the brightest stars are far hotter than the Sun. Despite this Svensmark and Calder believe that, when the biggest of the stars expire in mighty supernova explosions and spray the galaxy with cosmic rays, they do in fact cool the Earth by making the atmosphere cloudier.

When cosmic rays were detected by an Australian scientist nearly a century ago, it seemed that they were an interesting but unimportant extra, but it could be that they are an essential ingredient in the universe and a vital component in changing the climate on our planet.

Svensmark saw the first clues that cosmic rays have an effect on the climate when he looked at the alternating episodes of warmth and cold over the past few thousand years, starting with the Little Ice Age which peaked around 300 years ago, giving way to the present warm interlude. At the time of the Little Ice Age the Sun was in an unusual state, the Maunder Minimum, and there was very low Sun spot activity. This coupled with a jump in production rate of radiocarbon atoms and other long lived tracers, which are made by cosmic rays in nuclear reactions in the air, is an indication of low magnetic activity. Cosmic rays are deflected away from Earth by the Sun's magnetic field, but when it weakens more of them can reach the Earth. Since the most recent ice age ended 11,500 years ago, there have been nine chilling events like the Little

Ice Age and these have always been associated with high counts of radiocarbons and other tracers.

The cosmic rays must break through three defensive shields before they can reach the Earth's surface. First the Sun's magnetic field, then the Earth's magnetic field and finally the air around us. Only the most energetically charged particles can get as far as sea level. In Svensmark's theory it is these energetically charged particles called muons or heavy electrons, which are produced when cosmic rays hit the atmosphere, that help clouds to form low in the air and cool the Earth. Whilst some clouds higher up can have a warming effect; those clouds which are less than 3000 meters high keep the Earth cool. Put simply this means more cosmic rays, more clouds and cooler temperatures. During the 20th century the Sun's magnetic shield more than doubled in strength and so reduced the cosmic rays and clouds enough to explain a large fraction of the global warming reported by climate scientists. When they first revealed their ideas about the link between cosmic rays, clouds and the climate, Svensmark and his colleagues experienced a lot of criticism. To gain credibility for their theory, the team had to find out exactly how the cosmic rays affect the formation of clouds. Understanding of where clouds came from was surprisingly limited. Elementary text books said that when air becomes cold enough, moisture can condense and form clouds. But there must first be small specks floating in the air, the cloud condensation nuclei on which the water droplets can form. They needed to be seeded too, but how that happened was a mystery. The experiment SKY was set up in 2005 at the Danish National Space Centre and began to provide the scientists with some answers. Cosmic rays enter through the laboratory ceiling and into a large box of air, releasing electrons in the air which then encouraged the clumping of molecules to make micro-specks. The micro-specks are capable of gathering into the larger specks which are needed for the formation of clouds, the speed and efficiency at which the electrons worked took the team by surprise. In 2006 a more elaborate experiment CLOUD was set up at CERN, Europe's particle physics lab in Geneva, using accelerated particles to simulate the cosmic rays and test other possible effects. The influx of cosmic rays on the Earth depends not just on the state but on where we are in the galaxy. The Sun, along with the Earth, orbits round the centre of the Milky Way and sometimes finds itself in a dark region where hot, bright explosive stars are few. In those regions cosmic rays are relatively scarce and the Earth's climate is warm. This is referred to by geologists as the hothouse mode. In the opposite periods when the starlight and cosmic rays are abundant, the planet goes into an icehouse phase and ice sheets and glaciers form. An Israeli scientist has suggested that the major changes in theses phases could be accounted for by visits to the bright spiral arms of the Milky Way. (Svensmark, et al., 2007)

Although we do not fully understand the Sun-climate interface it is seems that Sun spot activity is a good proxy for solar activity in general. The 11 year Sun cycle is well known but other cycles of varying lengths have been suggested. One which is of particular interest is an approximately 1000 year cycle which is believed to have peaked recently. A previous peak in this cycle is reported to have produced the well documented medieval warming period. The forecast for the next two solar cycles, number 24 and particularly number 25, show that we are heading for a period of reduced activity for at least the next 15-20 years, perhaps until the mid-century and this would produce generally declining temperatures, as opposed to the temperature rises predicted by the IPCC. (Page, 29.11.07)

Recently Richard Black, the BBC environment correspondent wrote a report for BBC news online entitled 'No Sun link to climate change'. This article is about a new scientific study carried out by Dr Mike Lockwood of the UK's Rutherford-Appleton Laboratory and Dr Claus Froehlich from the World Radiation Centre in Switzerland. Their findings claim to disprove the cosmic ray hypothesis developed by, amongst others, Henrik Svensmark. The study was initiated by Dr Lockwood partly in response to the television documentary 'The Great Global Warming Swindle', which featured the cosmic ray hypothesis. He claims the "All the graphs they showed stopped in about 1980, and I knew why, because things diverge after that" "You can't just ignore bits of data that you don't like". That is simply not true. The graphs extend to at least 2000, as may be seen in the 2007 article in Astrophysics & Geophysics where at least four graphs go beyond 1980.

The scientists looked at the solar output and cosmic ray intensity over the last 30-40 years and compared those trends with the graph for global average surface temperature showing a rise of about 0.4°C over the period. This article does not go into a great deal of scientific detail to explain the findings of this new study, to support the claims there are two graphs, one showing the cosmic ray count and one showing the global mean surface air temperature from 1975 to 2005. As discussed before, the Sun varies on a cycle of approximately 11 years between periods of high and low activity, but that cycle coinciding with longer term trends saw most of the 20th century showing a slight but steady increase in solar output. All except for the period between 1985-90 when the trend appears to reverse and the solar output declines. However, the other graph shows that during this period the temperature rises just as fast. But although the graph does not show any obvious link, it is quite possible that it is part of a long term trend as was shown by Lassen and Friis-Christensen.

The study has been criticised by some for not recognising the work of Svensmark, Friis-Christensen and others. The article concludes that "changes in the Sun's output cannot be causing modern-day climate change". Lockwood does agree that there is a cosmic ray effect on cloud cover but thinks that, whilst it may have had a significant effect on the climate of preindustrial Britain, it cannot be applied today because the situation is completely different. Lockwood's analysis is said by Black to have "put a large, probably fatal nail" into the cosmic ray theory. (Black, 10.07.07)

In a follow up article, also entitled *'No Sun link' to climate change,* Black discusses the work done by scientists at Lancaster University to further contradict the cosmic ray theory. The team, headed by Professor Terry Sloan, has found that there has been no significant link between solar activity and cosmic ray intensity in the last 20 years. They presented their findings in the Institute of Physics journal and explained that they used three different methods to search for a correlation but found virtually none. To try and establish a link, Professor Sloan's team looked for periods in time and for places on Earth where weak or strong cosmic ray arrivals had been documented and then examined whether that had affected the cloudiness observed in those locations or at those times.

When speaking to the BBC news, Sloan explained that the Sun sometimes throws out a huge burst of charged particles, described as a 'burp', and that they had looked to see if cloud cover had increased after one of these bursts and that they found nothing. They did observe what they described as a 'weak' correlation between cosmic ray intensity and cloud cover over the course of one of the Sun's natural 11 year cycles, but concluded that, at most, cosmic ray variability could account for only a quarter of the changes in cloudiness.

Dr Giles Harrison of Reading University, who is a leading researcher in the physics of clouds, has also conducted research, looking at the UK only, which has suggested that cosmic rays only make a very weak contribution to cloud formation. Sloan says that the team did try to corroborate Svensmark's hypothesis but was unable to and the article concludes that the assessments made and the conclusions reached by the Intergovernmental Panel on Climate Change last year are correct and that Svensmark has no reason to challenge them. (Black, 03.04.08)

Conclusion

It is clear that there are many different theories for the cause of climate change and valid evidence to support them all. It seems that a great many politicians have made up their minds that greenhouse gases are the main driver of climate change, and they do not seem very willing to explore other possibilities. The media seems to have followed their lead. They report greenhouses gases being the cause of climate change as irrefutable fact and rarely report on any other possibilities. With sources of fossil fuels becoming more unstable because of dwindling stocks and political issues, warming due to carbon dioxide emissions gives even more reason to switch to alternate sources of energy, which as well as being beneficial to governments are also preferable to environmental campaigners. It has been suggested by some that politicians are providing scientists with funding to prove the CO_2 – temperature link and that the IPCC's findings are politically lead. Climate science is a big industry now with a lot of money invested and jobs depending on it.

By the end of the mid-century, the Earth is expected to have 3 billion more people to feed and, because CO_2 is the main source of plant food, the easiest and most environmentally harmless way to increase food production would be to double levels of CO_2.

This would have little effect on temperature (the CO_2-temperature forcing equation is logarithmic) but a significant effect on crop productivity, up to 40% in some varieties of soy bean. As the global temperatures fall, the oceans will absorb more CO_2 and CO_2 levels will begin to fall, this coupled with our efforts to try and cut CO_2 emissions will make world food production far more difficult. Because of this we must be very certain that CO_2 is causing a major change in the climate before we take action.

If cosmic rays are the main driver of climate change, it is good news for the world's inhabitants as it infers that the effect of carbon dioxide is quite small and, although there is nothing we can do about it, any global warming in the 21st century is likely to be much less than the typical predictions of 3 or 4°C.

It seems that the climate of our planet is such a complicated system that not enough is known to draw definite conclusions about the reasons for climate change. It does, however, seem difficult to believe that our species, that has dominated the planet for a relatively short period of time, could have such a huge impact on our planet's climate, whilst the Sun, the most massive body in the solar system whose influence dominates our planet, could have so little impact. This topic of the Sun's influence has received a significant boost from the recent publication of the book *The Neglected Sun*. This book was written by Fritz Vahrenholt and Sebastian Lüning and the significant boost mentioned is because the first author is a renowned German scientist, environmentalist, politician and industrialist who, amongst other things, served on the Sustainable Advisory Board for two German Chancellors – Gerhard Schroeder and Angela Merkel. In the book, the authors show that the crucial cause of global temperature change is the Sun's activity. They reveal that four concurrent solar cycles control the Earth's temperature and that this is a climatic reality with man's carbon emissions having little significance. The Sun's present cooling phase is considered in detail in this work and extreme doubt is cast, therefore, on recent catastrophic predictions emanating from the IPCC and the so-called 'green' agenda so prevalent in present day Western politics.

Clearly all the causes discussed here have an impact but their exact levels of contribution have yet to be determined. Possibly one useful step forward could be made by a reassessment of all the data gathered concerning the Sun's activity since the days of William Herschel. This, together with the data gathered in other countries over hundreds of years, needs to be re-examined before possibly erroneous claims are made which could have disastrous consequences for mankind.

References

"180 years of atmospheric CO2 gas analysis by chemical methods", Beck, Ernst-Georg ; Energy & Environment. -[s.l.] : Multi-Science Publishing Co Ltd, March 2007. -2 : Vol. 18.

An Inconvenient Truth , dir. Gore Al. -2007.

BBC Online News ; bbc.co.uk/science. -2008.

"Calculation of surface and top-of-atmosphere radiative fluxes from physical quantities based on ISCCP datasets: 2. Validation and first results" , Rossow, W.B and Zhang, Y.C ; Journal of Geophysical Research. -New York : [s.n.], January 1995. -Vol. 100.

"Climate change and Carbon dioxide , Page, Dr Norman J , Alpha Institute for Advanced Study. 27.06.07.

"Climate Change Science" Smith, Stewart ; Postnote (Parliamentary Office of Science and Technology). -November 2007. -Vol. Number 295.

"Cosmoclimatology: A new theory emerges", Svensmark, Henrik ; A&G News and reviews in Astronomy & Geophysics. -February 2007. -1 : Vol. 48.

email : Page Norman. -14.04.08.

email : Page Norman. -29.11.07.

Fourth Assessment Report-Climate Change 2007: Synthesis Report, IPCC Intergovernmental Panel on Climate Change. -2007.

"Global Warming? Its the coldest winter in decades", Bonnici, Tony , Daily Express. -18.02.08.

"Global Warming? New data shows the ice is back", Brennan, Phil , Newsmax.com. -19.02.08. 20 February 2008. - http://www.newsmax.com/printTemplate.html.

"Holocene thermal maximum in the western Arctic"; Kaufman, D. S., Ager, T. A. and Anderson, N. J., Quaternary Science Reviews, 2004.

"No Sun Link to Climate Change", Black, Richard, BBC News Online, 10.07.07.

"No Sun Link to Climate Change"; Black, Richard, BBC News Online, 03.04.08.

The Chilling Stars – A New Theory of Climate Change; Svensmark, Henrik and Calder, Nigel, Icon Books Ltd., 2007.

"The Great Global Warming Swindle", Durkin, Martin (dir), Wag TV, 2007.

The Rough Guide to Climate Change, Henson, Robert, London – Rough Guides Ltd, 2006.

"The temperature of Europe during the Holocene reconstructed from pollen data", Davis, B. A. S., Quaternary Science Review, 2003.

The Neglected Sun; Vahrenholt, Fritz and Lüning, Sebastian, Stacey International, London, 2012.

14. A Digression on Negative Temperatures.

Fairly recently, an extremely interesting article concerned with negative absolute temperatures appeared in the journal *Science*, entitled 'Negative absolute temperature for motional degrees of freedom'[1]. This article made no untoward statements or claims. Indeed, this article contained an exceptionally lucid account of the physics with which the authors were concerned. However, it was followed by an article in the journal *Nature* purporting to explain the aforementioned article in detail for the interested layman. In this latter article

http://www.nature.com/news/quantum-gas-goes-below-absolute-zero-1.12146

the author, Zeeya Merali, stated that the authors of the *Science* article had succeeded in cooling a system to a temperature *below* absolute zero and even went by the title 'Quantum gas goes below absolute zero'. It should be noted from the outset that this was definitely *not* what was claimed in the said *Science* article. The article actually claimed to have achieved negative temperatures in a system but with the concept of negative temperatures being defined in accordance with the accepted principles of thermodynamics; that is, where negative temperatures are *higher* than positive temperatures. This point was explained very carefully in the opening section of the *Science* article; there was *no* mention of any system achieving a temperature *below* absolute zero

The important difference reported by the *Science* article is that earlier experimental work usually involved examining spin systems, whereas this latest work seems to involve more directly understood physical systems. It should be noted also that, as well as misrepresenting what the authors of the *Science* article wrote, the apparent claims by Zeeya Merali do not accord with accepted thermodynamics as has been shown quite clearly. Hence, the importance of this present discussion. There are obviously people who claim to be scientists who do not understand this concept of negative absolute temperatures but write about them nevertheless. The issue has been raised yet again in the issue of *New Scientist* which came out on 22nd November 2014. This was in an article entitled 'Us versus the Universe' and, although the authors didn't explicitly refer to temperatures below absolute zero, they certainly wrote in such a way as to make readers consider such a happening. However, the journal itself published in quotation marks the statement that "A whole mirror-world of negative temperatures exists below absolute zero on the Kelvin scale". Nothing could be clearer than that scientifically incorrect statement. Obviously this is a dangerous state of affairs which this piece aims to address at least in part.

Negative absolute temperatures were first considered in the early 1950's via the work – both experimental and theoretical – of Pound, Purcell and Ramsey. All their work is well documented and the detail may be found in the listed references 2 – 5. However, it is worth noting at least some of the background as an introduction to the topic as a whole. As is indicated in references 2– 4, Pound,

Purcell and Ramsey examined various properties of the nuclear spin systems in a pure LiF crystal for which the spin lattice relaxation times were as large as 5 minutes at room temperature while the spin-spin relaxation time was less than 10-5 seconds. Various experiments were performed with the nuclear spin systems of this crystal, including some with a spin system at negative absolute temperature. Possibly the most important point to emerge here is that these negative absolute temperatures were achieved in physical systems in the laboratory. Admittedly the systems concerned were hardly everyday ones but, nevertheless, they were genuine physical systems. Obviously, if such systems did not exist in nature, there would be little, or no, point in studying negative absolute temperatures.

The detailed theoretical basis for this early work did not appear, however, until Ramsey's article of 1956 when most of the details were made known, including modifications, where necessary, to the wording of the laws of thermodynamics. The Clausius form of the Second Law remained unchanged but the Kelvin form had to be modified to

In a cyclic process, in the absence of other effects, heat cannot be converted completely into work for states of positive absolute temperature and work cannot be converted completely into heat for states of negative absolute temperature.

Again, the unattainability form of the Third Law had to be modified to

It is impossible in a finite number of steps to reduce any system to the absolute zero of positive temperature ($+0^0$K) or to raise any system to the absolute zero of negative temperature (-0^0K).

Various aspects of, and approaches to, thermodynamics make it seem an extremely abstract subject. Nevertheless, it is a branch of physics with roots firmly embedded in physical reality and whose purpose is to help in the explanation of physical phenomena. Nowhere is this link with reality better revealed than in the everyday notions of "hotter" and "colder". Here the everyday linguistic meaning of the terms is used in the physical theory. As Weinreich[6] points out, when two systems are placed in contact via a diathermic wall, the one which gives up heat is called the hotter and that which absorbs heat is the colder. The property of being hotter or colder is found to be transitive and this may be used to order all states of systems so that any state will give up heat only to states which are in lower positions on the list. The property determining position on this list is temperature and the hotter state is said to possess the higher temperature.

Again, each thermodynamic system must be capable of coming to thermal equilibrium with another system; that is, it must possess the property of thermal stability. This means that, if two systems at different temperatures exchange heat, the result must be to reduce the temperature difference between them. It follows from the First Law that if, in a process during which no work is done, heat flows from a hotter to a cooler system, the internal energy of the cooler system will increase while that of the hotter system will decrease. These changes must correspond to a warming up of the cooler system and a cooling down of the hotter system. This in turn implies that the temperature of each system must be a monotonically increasing function

$$\left(\frac{\partial T}{\partial U}\right)_{W=0} > 0, \tag{1}$$

where T and U represent temperature and internal energy respectively and $W = 0$ means that no work is done during the process.

The entropy S of a system may be written as a function $S(U, X_1, X_2, \dots)$ of the internal energy U and the deformation (or work) variables X_1, X_2,... Now, since

$$\left(\frac{\partial^2 S}{\partial U^2}\right)_{X_i} = -\frac{1}{T^2}\left(\frac{\partial T}{\partial U}\right)_{X_i} \tag{2}$$

where X_i indicates that all the X_i are held constant for these partial differentiations, the above criterion for thermal stability[6] implies that the curve of S against U is concave. Hence, if a system is capable of achieving both positive and negative absolute temperatures, the equilibrium curve of S as a function of U will possess a maximum and, for values of the internal energy less than that for which the maximum occurs, the temperature, given by

$$\frac{1}{T} = \left(\frac{\partial S}{\partial U}\right)_{X_i} \tag{3}$$

is seen to be positive; while for those greater than that for which the maximum occurs, the absolute temperature has a higher internal energy than an equilibrium state of positive absolute temperature at the same value of the entropy and work variables. Hence, in order to preserve the property of absolute temperature being a monotonically increasing function of the internal energy, negative absolute temperatures must be higher than positive absolute temperatures.

This latter point was emphasised first by Ramsey[5] who pointed out that, due to the form of the entropy curve discussed above for systems which exhibit both positive and negative absolute temperatures, it follows that, in cooling from negative to positive absolute temperatures, such a system passes through infinite absolute temperature and *not* through absolute zero. He also drew attention to the fact that the negative temperature cooling curves produced experimentally by Purcell and Pound[4] support this view. It is important to note that, once again, theory is supported by experiment and, therefore, any discussion of negative absolute temperatures and consequences of their existence is relevant to physics.

This may seem an unduly abstruse topic to consider in a book primarily designed to make members of the general public more aware of problems existing in the scientific world but which may impact, in some cases quite severely, on their daily lives. However, the story outlined above is highly relevant since it illustrates how that public cannot truly trust even the purportedly popular scientific magazines on sale at public news-stands. Further, it should be mentioned that this same misinformation is contained in the semi-popular book *The Laws of Thermodynamics: A Very Short Introduction*[7] and this is possibly even more serious as this book seems to be popular with students since it is very well and very clearly written but this, of course, renders it even more dangerous since impressionable young minds will be being filled with incorrect science.

It might be pointed out that both *Nature* and *New Scientist* were contacted over this matter. The person at *Nature* (whose name I forget), after some correspondence in which all the original basic references were cited, felt we had to 'agree to disagree' and there the matter ended. With due respect, one cannot really 'agree to disagree' on a matter of scientific fact. *New Scientist* didn't even reply to a letter. There the matter would have stayed had there not been a meeting with Professor U. M. Titulaer, who wrote the section on negative temperatures in the book *Problems in Thermodynamics and Statistical Physics* edited by Professor P. T. Landsberg[8]. He was both amazed and appalled that such incorrect information could appear in the two publications and not be rectified when the error was pointed out. It was only after that meeting that it was felt necessary to raise this matter again more publicly. Again, the members of the public, which ultimately pay the price of academic research, deserves to be kept aware of the whole picture on as many aspects of scientific endeavour as possible.

References.

[1] Braun, S., et al.,2013, Science **339**, 52

[2] Pound, R. V., 1951, Phys. Rev. **81,** 156

[3] Ramsey, N. F. and Pound, R. V., 1951, Phys. Rev. **81**, 278

[4] Purcell, E. M. and Pound, R. V., 1951, Phys. Rev. **81**, 279

[5] Ramsey, N. F., 1956, Phys. Rev. **103**, 20

[6]   Weinreich,   G.,1968,   *Fundamental Thermodynamics*, (Addison-Wesley, Reading, MA.)

[7] Atkins, P., 2010, *The Laws of Thermodynamics: A Very Short Introduction*, (O. U. P., Oxford)

[8] Landsberg, P. T., **1971, Problems in Thermodynamics and Statistical Physics,** (Pion. Ltd., London)

15. The Place of Mathematics in Physics.

Introduction.

During the last few decades, scientific experiments have increased in complexity. Hence, it has become increasingly important to attempt to explain some of the background to a wider audience, so that ultimately the members of the general public may have some real idea of the scope of the said projects.

Many look on in awe and wonder when told of the Large Hadron Collider. They have little idea what it is or, in reality, what those in charge hope it will do, but are carried along on a wave of quite probably genuine enthusiasm from those involved. The lack of knowledge, though, is emphasised by the genuine fear felt by some at the belief that, when switched on, this powerful machine would produce a black hole that would swallow up the Earth. Ridiculous as this may sound, there were people who did believe this and were genuinely stressed as the day of the switch-on drew nearer. The cost of this machine, as well as the enormous cost of running and maintaining it, are almost beyond the comprehension of many members of the general public. Then there is LISA, the Light Interferometer Space Antenna; another in that increasing group of projects separately costing vast quantities of public money. The question must be raised as to whether these projects ultimately gain the desired results.

There is little doubt that it would be extremely difficult, if not pointless, to explain the detailed thinking behind some of these modern projects in the general area of cosmology, for example, to the general public. This is not to appear élitist; it is rather that much of the theoretical background is so complex that relatively few professional scientists understand all the ramifications. Hence, how do you explain the background to people unused to the world of the professional scientist? It is not an easy task but is one that must be attempted and attempted with complete honesty. By honesty is meant the need to explain ALL the background. This would involve making everyone aware if alternative theories and explanations for effects and observations exist. At present, unfortunately, this is definitely not the case.

Discussion of the Basic Problem.

Much of the fear felt by so many as the day of the switch-on for the Large Hadron Collider approached was occasioned by a lack of knowledge of the real situation which arose for at least two reasons. Firstly, the explanations offered were necessarily sketchy because the concepts involved were so complicated and required vast amounts of background knowledge in physics to gain a true understanding. Secondly, however, no-one was made aware of the fact that other serious theories abound which made some of the worries pointless.

For over a hundred years now, scientific thought seems to have been held in the vicelike grip of two theories; - relativity and quantum mechanics. However, what of the qualms concerning those theories of relativity and quantum mechanics? It is well documented that many eminent scientists harboured doubts about the validity of relativity – both the special and general theories – from the beginning. Some, such as Herbert Dingle who became deeply troubled by aspects of the so-called twin paradox, formed doubts after initially being passionate advocates of the theory. Unfortunately, once those doubts arose, it seemed that eliminating them became increasingly difficult, if the account of events outlined in his book *Science at the Crossroads*[1] is accurate. Since those early days, little seems to have changed and, seemingly, it is still the case that challenging the validity of the theories of relativity is not a sensible career option. In fact, even showing that the famous tests of general relativity may be explained by other means[2] is regarded by some as a veiled attack on the validity of Einstein's theory, even when the author explicitly points out that no such interpretation is intended but that being able to tackle a problem from an alternative point of view can lead to greater clarity of the problem involved.

There have been worries expressed also over some points in quantum mechanics almost from the very beginning of the subject. Frequently, these have revolved around the role of the observer and over whether or not quantum mechanics is an objective theory. One man who has considered these points at length is Karl Popper, probably one of the best known philosophers of science. Although he has written on the topics at length, his book *Quantum Theory and the Schism in Physics*[3] proves an excellent source of his views. He expresses the view that the observer, or, as he prefers to call him, the experimentalist, plays exactly the same role in quantum mechanics as he does in classical physics; that is, he is there to test the theory. This, of course, is totally contrary to the so-called Copenhagen Interpretation, which provides the normally accepted position. This alternative view basically claims that "objective reality has evaporated" and "quantum mechanics does not represent particles, but rather our knowledge, our observations, or our consciousness, of particles". As Popper points out, there have been a great many very eminent physicists who, over the years, have switched allegiance from the pro-Copenhagen camp. He cites among these Louis de Broglie and his former pupil Jean-Pierre Vigier, Alfred Landé and, in some ways most importantly, David Bohm. Bohm, himself an acknowledged and deeply respected thinker, wrote a book on quantum theory, which was published in 1951, in which he presented the Copenhagen point of view in minute detail. Later, apparently under Einstein's influence, he arrived at a theory "whose logical consistency proved the falsity of the constantly repeated dogma that the quantum theory is 'complete' in the sense that it must prove incompatible with any more detailed theory". It was this very question of whether or not quantum mechanics is 'complete' which formed the basis of the intellectual struggle between Einstein and Bohr. Einstein said 'No'; Bohr claimed 'Yes'. The whole problem is discussed in great detail by Popper and, for those interested in this important topic, there can be no better reference than the book by Popper mentioned already. It should be noted also that people like Dingle and Bohm who have dared to question what might be termed conventional scientific wisdom have had their position within the scientific community brought into question.

Two enormously expensive undertakings mentioned earlier – the Large Hadron Collider and LISA – have much in common and illustrate well the need for increasing public understanding of some highly abstruse areas of modern science. Worries about the creation of black holes which could swallow the Earth troubled many. LISA will look for gravitational waves emanating from giant black holes. Hence, black holes are mentioned in both projects but what is the public's conception of a black hole and, indeed, of gravitational waves, and how was that conception achieved?

For many years now, black holes have been popular in science fiction and it is probable that, in many cases, the public's perception of what such an object is was derived from some work of science fiction rather than of pure science. This has been augmented by numerous television programmes, purportedly reporting genuine science. In truth, the programmes have reported science, but usually only advancing one explanation and ignoring other possibilities. The modern popular conception of a black hole is almost the perfect example of the public being misled as to scientific reality. Although the idea of a stellar body with an escape speed equal to, or greater than, the speed of light goes back to John Michell in 1784[4], the modern notion initially comes from Schwarzschild's solution[5] to the Einstein field equations of general relativity. There are at least two major problems associated with this and both are kept hidden from the public. Firstly, a simple check of Schwarzschild's original article shows immediately that the 'solution' so often quoted and used[6] is *not* actually Schwarzschild's solution. It is a later version due to someone else. The original does not include the mathematical singularity which leads to the idea of a black hole. Secondly, most modern work in this area of physics revolves around advancing explanations which depend on gravity only; the possible effects of any other forces are effectively ignored. However, most of the matter in the Universe is in the form of plasma. As such, electric currents will be circulating and magnetic fields will be playing a role. The electromagnetic force is much stronger than gravity by something of the order of thirty-nine orders of magnitude and there is a school of thought which feels that it is this force which plays the dominant role in the Universe, - not gravity! People advocating this alternative point out that black holes are simply not necessary in their scenario for describing the workings of the Universe. Incidentally, they also note that such esoteric notions as 'dark matter' and 'dark energy' are unnecessary as well. However, challenging the popular view is not allowed as it actually raises questions about the absolute validity of relativity and quantum mechanics. This means that the public, which ultimately foots the bill for all scientists do, is not being presented with all the facts before those very scientists embark on various extremely expensive projects.

It is an acknowledged fact that the well-known expert in English Literature, C. S. Lewis, claimed that everything he wrote was influenced by the Scot, George MacDonald – a man termed by no less a person than the poet W. H. Auden, to be 'one of the most remarkable writers of the nineteenth century'. Lewis compiled an anthology of readings from MacDonald's writings[7] and, in the present context, one of his comments on nature might be felt highly revealing –

'...the appearances of nature are the truths of nature, far deeper than any scientific discussions in and concerning them'.

The attitude reflected here is possibly illustrated further by another quote linked directly to botany in which Macdonald observes –

'To know a primrose is a higher thing than to know all the botany of it ...'

Many would do well to reflect on these two quotes.

Mathematics and Physics.

These latter two quotes, although not concerned with physics directly, highlight very precisely one of the major problems facing present day scientists in several distinct areas of the overall subject. It concerns the interface of theory and experiment/observation. In some older universities, mathematics as a subject for study was sometimes separated off from the so-called natural sciences and engineering and regarded as an arts topic with a B.A. degree awarded at the end of the undergraduate course. To many, mathematics is still not regarded as a scientific subject. From the point of view of the topic under discussion here, this is a highly interesting situation and one worthy of further contemplation. If mathematics is separated off from those natural sciences, one is left to consider why? What could be the reason for this? Any answer to such a question must necessarily be somewhat simplistic but suffice it to say that, in most respects, mathematics is a purely theoretical subject while in the natural sciences, specifically physics here, the emphasis is, or should be, more on observation and experiment. In the natural sciences, theory comes in when it is required to explain the results of experiments performed or observations made. It is, it seems, a quite natural human trait to wish to explain and understand what is going on around us, whether it be in our immediate vicinity or in the wider Universe. Anyone possessing an enquiring mind looking upwards on a clear starry night is bound to wonder about what he sees. What are those stars? Of what are they composed? How did they get there and how do they stay where they are? More detailed observations lead to even more questions. That is where theory comes in. People come up with ideas and need to write those ideas down in order to communicate them to others to obtain their views and hopefully finally come to an agreed answer to the original question. This is where mathematics crosses over to help out in the natural sciences. Mathematics is the language used to write down the thoughts developed to answer specific questions in physics – be it in astronomy or any other branch of the overall subject. Hence mathematics enters the picture as an aid to the natural scientist; mathematics is a tool to aid in the final formulation of a piece of theory. Note that this means that the mathematics is, in a very real sense, subservient to the original science. The test of the model is then to see how well it explains the original observations/experiments. If it passes that test then, and only then, does it become permissible to see what predictions the model may make. Any such predictions must then be tested rigorously by further observations/experiments but at no time should anyone be tempted to even think of making the observations or experimental results fit the theory; the opposite approach is the only sensible and correct way forward, that is, to check that the theoretical predictions are valid. Again, at no time should anyone be tempted to assume the theory provides the ultimate answer to the branch of science under consideration. As quoted earlier from Lewis's anthology of MacDonald's writings

'To know a primrose is a higher thing than to know the botany of it...'

Here, for example, the 'primrose' might be a star or a galaxy and the 'botany' would be the physical model involved.

The twice quoted remark from MacDonald provides a truly sobering thought but one which all scientists would do well to read, contemplate and digest. It should make all realise that we are all floundering around attempting to solve a huge problem – trying to devise a theory or theories to explain all we see and experience in the world, even the Universe, around us. Such a thought should, if absorbed properly and completely, have a profound effect on all and make all a little more humble and willing to explain things as simply, accurately and patiently as possible – and that may not always be an easy task – to the public at large. Truth should be all important but also a realisation that mathematical models are not the whole truth, merely approximations to that desired truth, must be at the forefront of any explanations.

Concluding Remarks.

It is interesting to note that, in a recent issue of the journal Astronomy[8], an article appeared discussing the position of mathematics in astronomy. Some of the answers given by professional astronomers were particularly intriguing with one even claiming that 'as an astronomer, maths. is all I do'. Another pointed out that 'an investigation into the nature of astronomical objects requires understanding the underlying physics, which therefore involves maths.' There may well be some truth in the second quote but it promotes a potentially dangerous notion. There is surely no immediately obvious reason for mathematics to play any role in understanding some underlying physics. It may be true in some cases but the real understanding should normally come from the physics itself by way of observation and experiment. This was certainly true with the work of Birkeland to whom reference was made earlier and a perusal of his life[9] shows that much the same could be said of Newton who is probably regarded today as the greatest of theoreticians. In his day, Newton based virtually everything on observation and experiment and it was those two areas, subsequently analysed by one of the most perceptive brains ever, which helped lead to his major theories which exist almost unchallenged to this day. It seems that many, possibly most, of the truly great advances of the past were made with mathematics as the tool introduced to aid the investigation but the main driving force seems to have been the physics, that is, the observation and experimentation involved. In the case of Newton, one of his massive intellectual achievements was the development of completely new branches of mathematics to aid him in his overall work but, at all times, it was the physics – the observation and experimentation – that provided the main driving force. This vitally important point is emphasised beautifully by Tait[10] in his inimitable style when he notes that 'In dealing with physical science it is absolutely necessary to keep well in view the all-important principle that

Nothing can be learned as to the physical world save by observation and

experiment, or by mathematical deductions from data so obtained.'

Today's scientists could do worse than remember this and take the idea to heart. They might benefit also from reading and studying at least some chapters of that intriguing book on Newton – *The Cambridge Companion to Newton*[11]. In the present context it is illuminating to note some of I. Bernard Cohen's comments in chapter 2 –

'Newton's goal is eventually to get to the dynamics of the system of the world. But he makes it abundantly clear that in Book 1 he is primarily concerned with elaborating the properties of mathematical systems that have features resembling those found in nature. (Although stated quite clearly, this was incidentally something not accepted by some, especially on the continent of Europe.) And here he makes an important distinction between mathematics and physics. In this way, Newton is free to develop the properties of mathematical forces of attraction without having to face the great problems of whether such forces can actually exist or can be considered an element of acceptable physics....As Newton proceeds step-by-step, he introduces into the mathematical system one-by-one such further properties as will make the system more and more closely resemble what we observe in the world on nature.' Hence, again as observed by Cohen, the essence of the Newtonian Style is this notion of adding the conditions resembling those of our world one by one; that is, increasing the complexity and accuracy of the model bit by bit but always noting the physics of the situation as being all-important. It does seem that, if such an approach was good enough for Newton, it should be the path followed by today's scientists, not just in physics but probably in all areas of science.

Mathematics is undoubtedly a very beautiful subject in its own right and is a worthwhile intellectual exercise for anyone to study. Also, research in mathematics for its own sake should be encouraged but, when mathematics is used to aid in the solution of a problem in physics, it becomes a mere tool to help the investigator.

A further related point is that science should be studied with a totally open mind, as certainly seems to have been the case with Newton, and any advances should be examined in a like manner. Surely the aim of any scientific investigation is to seek the truth? Probably mankind will always be found wanting intellectually and any solution to a problem will be no more than an approximation to the real truth, but efforts must continue in all areas to find that elusive complete answer. In the meantime, the dissemination of scientific information to the public must be totally honest and open. Where several theories exist, that fact must be openly acknowledged with no thought for protecting vested interest of any sort. The task will be extremely difficult because of the nature of the technical language and theory involved but it must be attempted and attempted by genuine scientists involved in the work rather than pseudo-scientists who happen to be good professional presenters.

References.

1. H. Dingle, *Science at the Crossroads,* Martin Brian & O'Keefe, London (1972)

2. B. H. Lavenda, J. App. Sc. 5, 299 – 308 (2005)

3. K. R. Popper, *Quantum theory and the Schism in Physics*, Hutchinson, London (1982)

4. J. Michell, Phil. Trans. R. Soc. 74, 35 (1784)

5. K. Schwarzschild, Sitzungsberichte der Königlich Preussischen Akademie der Wissenschaften zu Berlin, Phys-Math. Klasse, 189 (1916)

6. see for example:
    R. Adler, M. Bazin, M. Schiffer, *Introduction to General Relativity*, McGraw-Hill, New York (1965)

7. C. S. Lewis, *George MacDonald, An Anthology – 365 Readings*, HarperCollins, San Franciso (2001)

8. R. Berman, Astronomy, December, 24 (2013)

9. See for example:
G. E. Chistianson, *In the Presence of the Creator,* Macmillan, New York (1984)

10. P. G. Tait, *Recent Advances in Physical Science with a Special Lecture on Force,* Macmillan & Co., London (1885)

11. I. Bernard Cohen & George E. Smith, *The Cambridge Companion to Newton,* Cambridge U. P., Cambridge (2002)

16. The Public Funding of Science.

In any society, whether democratic or not, the individual must necessarily have certain bounds placed on his economic freedom by the obvious necessity of the prevailing government to raise taxes to pay for the various services we and the country need. In most cases, although the individual may not know the precise details of the way in which collected revenue is spent in particular areas, he will have some rough idea of the need for some expenses. Although some might disagree, most would see a need to spend money on defence; most would agree with funding education in general terms; most would agree to spending on health and social requirements; and so on. In each case mentioned, there could well be arguments over the amounts spent in specific areas and on specific projects but, in general terms, any government would have broad agreement to spend in these areas. Of course, each of these mentioned areas and others which have not been cited benefit from the fact that all have a direct effect on all individuals in the state and all feel, rightly or wrongly, that they know something about these areas. Hence, all are able to form opinions on whether or not a particular piece of expenditure is correct. The individual will base his view on information gathered in the main from various branches of the media, together with possibly a little background reading and discussions with friends and colleagues. The view expressed finally may not be truly well informed but at least the individual feels he can have a sensible say. However, is this the position when one comes to discuss the funding of science? This is undoubtedly a much bigger question than many realise and is becoming increasingly important in these days when so many hugely expensive science projects are being proposed and funded.

The views expressed by the scientific laymen are formed once again by interaction with the various arms of the media. These days the brilliantly produced science programmes on television are hard to resist and the message they send even harder. Technology is used to great effect and one can well understand how the young and the scientifically uninitiated are left impressed and, more importantly, convinced by the arguments they've had thrust upon them in their living rooms. However, are these members of the public given the true overall picture? Are they made aware of conflicting views, if such exist? Are they made aware of alternative theories, again if such exist? The answer to all these questions is 'No' and this is a grave concern for many people in the know, both scientists and non-scientists. Further it is worrying to realise that it is quite possible that those in the position to make the final decision on the allocation of scientific research funding are also not necessarily in possession of all the facts when they have to make specific decisions.

The above is a serious allegation to make but is supported by an extremely strong body of evidence, but evidence of a type that will only be viewed sympathetically, let alone accepted, by people with open minds willing to question authority in a mild, nonaggressive manner. The cases to be used to illustrate this assertion will be taken from the general area of physics in the main but are all examples well-known to the lay public due to their constant coverage in the media and in so many popular science books.

Einstein's theories of relativity.

In the nineteenth century, the existence of a material medium, the æther, pervading all space was a generally accepted concept. The supposed mechanical vibrations of this medium were used to explain the wave propagation of light. One great challenge facing experimentalists, therefore, was to detect the actual presence of this medium. At the time, optical experiments were the most accurate available. Easily the best known was that performed by Michelson and Morley in the 1880's. It is well recorded that this experiment failed to detect the physical existence of the æther. In the history of the development of special relativity, this is the first juncture where questions should be raised. Was it actually true that the experiment did fail to detect the physical existence of an æther? The controversy surrounding this straightforward question continued throughout the twentieth century and is not resolved even today. It is claimed in the vast majority of, if not all, textbooks that no absolute motion was detected but, in truth, the published data revealed a speed of 8km/s. However, this made use of Newtonian theory to calibrate the equipment and was a figure much less than the 30km/s orbital speed of the earth. It was purely due to this second point that the detected speed was less than the orbital speed of the earth that a null result was claimed. It is now claimed by some that modern analysis leads to a different calibration for the equipment and that this, in turn, leads to a speed in excess of 300km/s. The claim is then that the experiment both detected absolute motion and the breakdown of Newtonian theory. This first supposed detection of absolute motion has supposedly been confirmed by other experiments.

However, it quickly became accepted generally that the Michelson and Morley experiment did, in fact, fail to detect the existence of an æther and there then resulted a major challenge to the theoreticians to explain this null result. After much preliminary work by such as Lorentz and Poincaré, Einstein's special theory of relativity emerged as the accepted explanation although it should be realised that most of the results had been produced by Poincaré up to eight years earlier. Also, that most famous of equations, $E = mc^2$, had been known and used for several years before Einstein's work was published. As one example to support this assertion, this particular result was mentioned in Thomson's book *Electricity and Matter*, which appeared in 1904 but was actually the text of lectures delivered at Yale University in 1903, but the result's history goes back much further than that. However, be that as it may, since those early years of the twentieth century, there has been much discussion of the results of the Michelson-Morley experiment; it being claimed on many occasions that the experiment did not, in fact, produce a null result. The controversy still exists.

It should never be forgotten that Einstein also thought very deeply about the problem of gravitation. Whether or not he turned his attention to this because of a problem with the orbit of the planet Mercury is not really important, although it does provide a convenient starting point for any discussion of what is now known as Einstein's General Theory of Relativity. The name merely indicates a follow-on from his special theory but, in fact, it is really a theory of gravitation although, like all other theories of gravitation, it doesn't explain exactly what the force of gravity really is. The final point is not at all surprising since no-one really understands what a force is, merely what it does! It is often pointed out that people such as Poincaré and Lorentz contributed greatly to the special theory of relativity but, where the general theory is concerned, the tremendous intellectual achievement was Einstein's own. True he made use of the mathematical results of such as Riemann, Bianchi and Ricci, but the final physical theory was entirely the work of Einstein himself; he merely made use of known results in differential geometry in the same way as others utilised known results in algebra or calculus. As well as explaining the problem associated with the orbit of Mercury, the theory also made predictions concerning the bending of light rays as they passed a massive body such as the sun. This offered almost immediate scope for scientists to test this revolutionary new theory. The eclipse of 1919 provided the perfect opportunity. The observations made of this eclipse by Eddington were used to herald the almost complete vindication of this theory, although subsequently doubts have been cast over the actual information obtained at that time. Incidentally, according to Herbert Dingle, as recorded in his book *Science at the Crossroads* (1972, Martin Brian & O'Keefe,

London), it was only after Eddington's apparent vindication of his General Theory of Relativity that Einstein's Special Theory assumed precedence over the earlier theory advanced by Poincaré, - a theory which, incidentally, incorporated an æther.

Hence both Einstein's theories of relativity, while forming the basis for so much modern research in physics and astronomy, still have genuine question marks hanging over their validity. However, while many people have heard of Einstein's theories, not many have real knowledge of them. That is not the case with two topics which have arisen out of those theories and have been very much brought to the public's attention via the media in its various forms. These topics are the idea of a Big Bang as the origin of our Universe and the notion of black holes. Both topics have been covered in great detail in serious television science programmes, in many popular science books as well as in numerous science fiction books aimed at people of all ages. The result is that many people feel they really do know something about these two topics, but do they?

The Big Bang.

The whole idea of the Big Bang as the starting point for our Universe goes back to the theoretical work of Alexander Friedmann and Georges Lemaître in the earlier years of the last century following publication of Einstein's General Theory of Relativity. Its movement to a position of prominence, if not pre-eminence, in cosmology might be felt to have been brought about by its eloquent advocacy at the hands of George Gamow in the mid and late 1940's, ably supported by such as J. Robert Oppenheimer. Of course, if anyone dares ask the seemingly childlike question 'What went bang?', confusion tends to follow.

However, the Big Bang as a valid model of the Universe has been under close scrutiny almost since it was proposed and many of the queries concerning it remain. These queries tend to be 'swept under the carpet' but in a rather subtle way. The rise of popular science books has provided a means whereby the general public is persuaded to believe in the ideas accepted as founding 'conventional wisdom'. The 'solutions' to various problems are presented as indisputable facts; the notion that other possible explanations exist is carefully suppressed. One notable exception to this observation, although it probably wouldn't be considered a 'popular' science book, is the *Cambridge Encyclopædia of Astronomy*, which appeared in 1977. This book provides an excellent example of a book which, while apparently supporting the commonly accepted view of things, nevertheless obviously leaves the door open for other explanations of observed phenomena. In many ways, it provides an object lesson in open-minded, unbiased writing of a scientific text - popular or otherwise.

However, almost from the beginning, a problem faced by the adherents to the theory was, and still is, the seemingly constant need to add to the basic theory in order to overcome problems. Obvious examples of this are the introduction of the ideas of inflation, dark matter and even dark energy, more abstract notions with which most are only too familiar if only by name. It is not, I think, without interest to realise that additions to the Big Bang theory such as those mentioned are accepted unerringly. Seemingly, no questions are raised when these new notions are introduced in attempts to preserve this theory as the only acceptable explanation for our universe as we see it.

Unfortunately, it is only too true that, at the end of their undergraduate days, many students emerge totally convinced that the Big Bang theory correctly describes the beginnings of our universe and also many of its subsequently developed properties. They believe it to be the only theory which explains the cosmic microwave background radiation; they believe it to be the only theory to explain the mass fraction of helium. This, and much more, has all been learnt in undergraduate courses as being absolutely sacrosanct. Further, these beliefs are vigorously supported by so many popular science books, such as Simon Singh's *Big Bang*, and by many popular science lectures. The popular science lecture on the Big Bang by Simon Singh, which has received quite widespread publicity, is an excellent example. This lecture is beautifully constructed and presented, as one might expect from an experienced member of the BBC personnel able to call on the resources of that organisation if necessary. The personality presenting the talk is friendly and engaging; a young audience, in particular, is rapidly enthralled. The use of power point to deliver the message, and message it is, is very professional. Everything about the talk from a delivery point of view is beyond reproach, and that is where the danger lies. Young people with impressionable minds will leave such a talk totally convinced that they have just been exposed to an enunciation of the complete truth regarding the birth of our universe. But have they? They will have been told, amongst other things, that the cosmic background radiation was discovered by Penzias and Wilson in 1965; the actual published discovery in 1941 by McKellar will have been ignored. The Steady State theory will have been dismissed totally with hardly a glance in its direction and no mention will have been made of the newer modified theory. The constant need to add to, and modify, the original Big Bang theory will have been glossed over. However, in the lecture being highlighted here, the presentation will have been so slick and professional that these points will not have sunk in to members of the audience. Many of the enthralled youngsters will probably leave the lecture theatre remembering more that Simon Singh would like to be admired by Cameron Diaz in the same way that Einstein was apparently admired by Tallulah Bankhead, than that they have just heard details of *one theory* for the beginning of our universe. Superficial gloss will have prevailed. As stated previously, herein lies the danger. The scientists of tomorrow are not being trained to have open questioning minds. Rather they are having their minds programmed to be closed to all thoughts which might possibly conflict with 'conventional wisdom'. The message often appears to be delivered with what amounts to an almost religious fervour, – what might be termed scientific evangelism.

Comment must be made at this juncture about the latest addition to this field of scientific indoctrination. Following last year's successful series *Wonders of the Solar System,* the nation is to be treated to the same presenter pontificating on the *Wonders of the Universe*, and this by a person who is not an astronomer or astrophysicist. This latter point is important because it means the material being presented will be prepared by someone else and one can almost guarantee it will be biased in favour of the prevailing status quo. I have little doubt the programmes will be good television and will attract good audience figures but I also fear for the content. Once again many of those in the audience will be highly impressionable youngsters ripe for glamorous indoctrination and that, I strongly suspect, is what will happen. Of course, many older people will also fall for the 'boyish charm' and be hoodwinked into believing that they are hearing the actual facts of the situation. Few will realise they are simply hearing about one theory and one with many questions hanging over it – questions which I fear will go unmentioned and definitely unanswered. These questions range from 'Why the need to attach so many additions to the original theory?' to 'Are the alleged predictions of the theory peculiar to this one theory?' In truth, the answer to the first question is that these additions are necessary to rescue the theory from total rejection. The answer to the second question is, quite simply, No! For example, the temperature of the so-called cosmic background radiation is found quite accurately by several other theories, including the Steady State theory and it must always be remembered that the so-called cosmic background radiation itself admits several explanations for its existence. Hence, the question of the validity of the Big Bang theory remains an open one.

Black holes.

Much the same story holds true for black holes, those peculiar stellar bodies so beloved of science fiction writers. Popularly, a black hole is taken to be a body so dense that one would need to move at a speed equal to, or greater than, that of light in order to escape from it. A body with this particular property was discussed in 1784 by an Englishman named John Michell using ordinary Newtonian methods and the type of body he was considering could exist theoretically. However, the modern notion of a black hole is somewhat different and results from an attempt to give a physical explanation to a mathematical singularity which crops up in general relativity. For the uninitiated, a mathematical singularity is a point where a quantity takes on an infinite value. Years ago such a point used to be thought a point at which the theory under consideration broke down but nowadays some people adopt a different stance. Over the years, a great many objections have been raised to this concept. Many focus on the mathematics involved and indicate where incorrect steps appear to have been taken. Nevertheless, popular opinion demands the possible existence of such bodies and we are informed quite regularly that yet another black hole has been found or that all galaxies have at least one black hole at their centre. In fact, as yet, no object has been identified as a black hole beyond reasonable doubt.

However, the myth remains and it seems that, in the public eye, because of the information given to members of that public and the manner in which such information has been disseminated, black holes are not mythical in any way; they are real!

Deductions following.

The above are simply three examples of scientific theories which have been presented to the general public in a very popular, highly biased way. They have, however, been presented virtually as accepted fact, rather than as mere theories whose actual validity is far from established. The end result of this is that, when applications for large sums of money to fund research projects such as the Large Hadron Collider, LISA and LIGO are presented, because they are said to rely on Einstein's theories, may help establish the big bang model even more firmly, and may even produce some mini black holes, in general the public has few qualms over such expenditure. Further, the uninitiated probably feel that everything has been through the so-called peer review process and so the whole procedure must be completely satisfactory. However, just how true is this? How reliable is this peer review system?

Peer review.

Firstly one must ask the obvious question of what is peer review precisely? Somewhat surprisingly this is not a straightforward question to answer even though it is a process at the very heart of the operation of academic journals and of grant awarding. Clearly it has to do with some third party reviewing an article prior to possible publication or reviewing a grant application for the awarding body. Normally, it would be expected that the reviewer would be an expert in the field covered by the article or grant application but, particularly in highly specialised cases, this may not always be so. Hence, it may be wondered as to the meaning of the word 'peer' here. Again, people may wonder how many reviewers there should be in any particular case; they may wonder whether the authors' names should be with-held from the reviewer, although this is rarely the case; they may wonder if the reviewer should be anonymous as is usually the case. These are just a few of the questions which may be raised in connection with this process which is seemingly at the heart of so much in science and can have such a dramatic effect on the lives and careers of the researchers involved as well as a serious, if indirect, effect on everyone. This final remark relates in particular to the area of medical research where, if someone makes an incorrect choice or decision, it can lead to human tragedy. However, in other areas of science, it can also lead to effects which have dire consequences – often financial – for the general public.

Over the years many have speculated on both the effectiveness and fairness of this system which almost seems to be a part of the foundations of scientific research. Whether or not detailed examinations have taken place in other fields, it is certainly the case that there has been much work examining this topic in medicine and the results are worrying if not truly unexpected. One editor of an eminent medical journal wrote that 'if peer review was a drug it would never be allowed onto the market' because 'we have no convincing evidence of its benefits but a lot of evidence of its flaws'.

Of course, here attention is restricted to the peer review process utilised to accept/reject articles for publication or to decide on who does and does not receive research funding. Once an article appears in print or on the internet, it is subject to detailed scrutiny by all who read it and, to many, this is felt a far more rigorous and fair form of peer review; rigorous because the work is examined by far more people; fair because people are, at that stage, not involved with protecting their own, or friends', reputations. Here, unfortunately, it has to be realised that science is no pure search for scientific truth; there is a huge amount of protection of personal position and of what might be termed 'conventional wisdom'. The three topics in physics mentioned earlier all fall into this category of conventional wisdom. No-one dare publicly challenge the validity of Einstein's theories of relativity; the big bang model for the beginning and development of our Universe is almost sacrosanct; and the actual presence of black holes is readily asserted with no fear of rejection even though the theoretical evidence from which their presence is deduced is dubious to say the least and no such object has been identified beyond all reasonable doubt – contrary to so many printed reports. Hence, a review method which many in medicine have found to be seriously flawed is still in place in all science and is undoubtedly holding back advance. This is, quite simply, because, as pointed out with the three instances above, certain topics are simply not open to investigation; the 'gods' ruling science have decreed for reasons best known to themselves, but probably connected with self-aggrandisement, that such topics are now closed for discussion and further examination.

Of course, it is easy to criticise but less so to offer a viable acceptable alternative which is also an improvement on the present system. As far as the assessment of academic articles is concerned, it has been suggested quite seriously that all peer review could, and possibly should, be abolished and each individual piece of work examined and rated by the entire scientific community. This would not have been practical in the days when only paper journals were available but nowadays, with the internet it does become a viable alternative. It has been tried with a number of freely available sites but with varying degrees of success. It has to be admitted that people have begun to impose restrictions on some of these sites when they've been operating for a short while. The problem with this is that the restrictions have usually been introduced to prevent the appearance of some points of view since, obviously, the amount of space available is not a problem with an online publication. This is what has happened with the arxiv site administered by Cornell University. This site set out to be freely available to all to post articles, but restrictions have been imposed more and more in recent years and it has now reached the stage where experts are being denied the right to respond to criticisms of well-established theory in their field by relatively unknown people. The end result of this, though, may be beneficial to science in the long run in that it has lead to the establishment of alternative on-line sites, such as vixra, to combat this censorship being imposed by a seemingly self-appointed clique.

However, if an online journal starts out by being available to all, so long as it remains completely open, it seems to offer a possible solution to one problem with peer review. However, what of the problem where grant applications are involved?

Here the problem is entirely different and it is one which assumes added importance with the announcement that the present dire economic conditions require a further concentration of research funding on top-rated work. However, what is top-rated work and who defines it? It must be realised that most work that is truly top-rate can become classified as such only *after* its completion. Again, whenever, money is involved, people always have pet projects which they feel must be funded before all else. For example, whenever there is talk of cutting science research funding, some body of people will immediately start proclaiming the vital importance of their work for mankind and claiming that, although they recognise the need for saving, their area must remain virtually untouched because of the possible benefits for all that could result from their work. While such an argument may have some merit in some fields of medical, or medically related, research for example, the devastating benefits for mankind arising out of a new telescope or particle accelerator being built somewhere are not so immediately obvious. Nevertheless, for this second example cited, how is a fair and just decision reached? Until now, peer review has provided the answer but how fair and just has it been? It must be acknowledged that no human system will ever be perfect but, in this area, the system developed does seem far less than perfect. The individuals acting to review applications are human beings and, as such, susceptible to the failings of all human beings. Many would feel that members of these review panels favour not just their own disciplines but their own particular speciality within that discipline. Is this a fair point? Possibly not in some cases but, over the years, the system has produced so much discontent that feelings of injustice abound. The stories from colleagues are legion but, to give just one example to illustrate the point, it cannot be acceptable to totally reject a proposal purely because the applicants haven't published in the precise field of the application before. This, however, can be the case.

Here though abandoning peer review is not so easy. One point that is forgotten often is that, where grant applications are concerned, there will only be a finite amount of money available and so some applications will necessarily be successful, others not. How, therefore, is the available money to be apportioned? It seems that a version of the present peer review system must remain but, from what has happened in more recent times at least, it seems that some safeguards must be introduced. Here it seems that the availability of more and fairer knowledge of what is going on in science might provide a good starting point. Apart from the professionals who carry out the peer review, the process will necessarily involve input from non-scientists, for example, of some civil servants. It might be useful if these people were more aware of the real truth of some areas of science. Here the three topics to which I alluded earlier provide excellent examples of areas where the public has been hoodwinked, at least to some extent. People have been placed on academic pedestals and, once that has happened, from that moment on they and their work simply cannot be challenged. True science should involve a genuine search for the truth about a topic or area and, as such, views and ideas may well change over the course of time as instrumentation becomes more and more accurate and as new instrumentation becomes available. Science cannot, by its very nature, stand still.

The validity of Einstein's theories of relativity has been challenged almost since they were first enunciated. Alternative solutions to the various problems those theories were supposedly developed to examine have been proposed on numerous occasions but have been rejected, not because they were proved incorrect but because they were thought to challenge the validity of relativity. In truth, all most of these pieces of work did was offer an alternative solution to a problem. No more, no less!

The Big Bang is simply a theory of how our Universe came into being and developed but that is all it is – a theory! As such, it should be challenged and any challenge should be taken seriously, not dismissed simply because it queries conventional wisdom. Black holes are merely theoretical constructs but the public has been led to believe the existence of such objects is established fact. On the other hand, the public is largely unaware of alternative explanations in existence for these phenomena. Many might have heard of the so-called Steady State theory for explaining our universe. However, most feel it has been successfully discredited, but has it? The answer to this question is definitely 'No'. True, there are problems with this theory as with all theories but, in its present form, it can describe accurately the phenomena concerned at least as well as the big bang theory. However, both these theories rely totally on the force of gravity to explain things.

Most of the matter in our Universe is electrically charged, being in the form of plasma, and the electromagnetic force is 39 orders of magnitude greater than the force of gravity; that is, you multiply the magnitude of the force of gravity by one followed by 39 zeros to find the corresponding magnitude of the electromagnetic force. Plasma has been studied in laboratories for in excess of one hundred years and the scientists concerned, including Nobel Prize Winners, have produced effects reminiscent of astronomical phenomena on numerous occasions; for example, an effect similar to the aurora borealis has been produced. Bringing the effects of electricity and magnetism to the fore, as plasma physicists have attempted to do, has produced many new – experimentally backed – explanations for many astronomical phenomena. These are not always welcome developments though since they have been accompanied on occasions by new ideas about some of the heavenly bodies. Looking at things from the plasma point of view brings different models of the stars into the picture. However, where our Sun is concerned, this has meant the emergence of explanations for phenomena which had, and still are, puzzling many astronomers. No; these alternative ideas have not been accepted; rather they have been ignored!

This then is the background to the very real worries expressed about the huge expenditure of public money on some of these vastly expensive projects such as the large hadron collider. If money was not a problem, no-one could really harbour objections to projects such as these but, especially in these days of belt tightening for the man in the street, should some privileged scientists be allowed these excessively expensive toys?

Conclusion.

A large proportion of the funding for scientific research in the West ultimately comes from the public purse. The ordinary man-in-the-street is the one who, in the final analysis, pays for much of this research through the taxes collected. At present, although he may wonder at the reasons for some areas of investigation, when it comes to many of the hugely expensive projects, he has been lulled into thinking financial support is being given to some really worthwhile fundamental work based on solid theoretical foundations. This is not the precise language that might be used to describe the situation but it does convey the precise sentiment involved. Unfortunately, this is not an accurate picture of the situation facing the public.

Today, with the advances in all forms of communication, cult status has been seen to have been afforded to so many who, in a bygone age, might well have remained in deserved obscurity. Such people are deemed to possess charisma. As mentioned earlier in connection with the so-called Big Bang theory, George Gamow's eloquent advocacy of this theory earned him a type of cult status which, as a theoretical scientist, he might not have obtained otherwise. Here is seen an early example of the public being fascinated by a piece of abstract scientific theory and the purveyor of this information gaining publicity probably beyond his wildest dreams. However, here, Gamow was a genuine advocate of a theory on which he was working and to which he was producing original contributions. Nowadays it seems the purveyors are, in reality, professional purveyors of information; most are not scientists who are themselves working at the boundaries of knowledge. These people are, though, totally professional in their job and, as such, might be complemented. However, they are usually purveying ideas communicated to them by interested parties. By interested parties is meant parties whose overriding interest is in their own ideas and beliefs being afforded as much positive publicity as possible. This, in itself, virtually guarantees a balanced view of a topic being ruled out. An added point here is that, nowadays, attracting money to a university is often a far more important factor affecting a person's promotion than what is actually achieved research-wise with that money. It used to be said that 'money is the root of all evil' and, in present day science that seems an accurate statement. As a small aside, it's interesting how so many of these old sayings appear to have disappeared from our everyday language. In so many ways that is a great pity because so many are so apt and so accurate on so many occasions.

However, to return to the public funding of science; it has surely been noted by most people that the financial demands of the science sector, both inside and outside our universities, are steadily becoming greater and greater. Whenever, restraint is urged, as in our present dire financial times, everyone seems to agree that such restraint is necessary but, unfortunately, everyone also seems to agree that such restraint cannot possibly occur in their particular sector. However, even if there was a pot of gold at the end of the rainbow, more true accountability when it comes to funding scientific research should be involved. Too often nowadays people openly say that results have been obtained because they were necessary to ensure continued funding. It has been said on numerous occasions that, in applying for a particular research grant, it was implied that certain results would emerge and so such results had to be obtained to ensure continued credibility for those involved and, hence, continued funding. Are these stories merely apocryphal? Who knows? However, to a great extent, the truth or otherwise of these stories is immaterial. The terrible truth is that they exist at all. The origin of such stories cannot be simply put down to 'sour grapes'. At least some feel them to be true and that only casts doubt on scientific claims in general. Added to this, the public has heard of the well-publicised cases of scientific fraud where researchers have claimed results that were simply not true but have had the dubious work accepted for publication in prestigious peer-reviewed journals long before the fraud was exposed. There have also been cases where referees have prevented publication of an article, only to steal the results and try to publish them themselves. One classic case is mentioned in the important article on peer review in medicine (Richard Smith; 'Classical peer review: an empty gun'; http://breast-cancer-research.com/content/12/S4/S13) but one wonders how many similar cases have slipped through undetected. It might be noted that the case mentioned came to light simply because the final article was sent for refereeing to someone in the same department as the author of the original rejected article. Hence, depending on your point of view, it came to light purely because of a lucky/unlucky coincidence.

However, one would strongly suspect this final example a rare occurrence, especially when compared with a personal need to preserve a reputation and position. Very often, unfortunately, people have built extremely successful careers by selling their souls to a particular theory. In many ways, this is entirely understandable but it does demonstrate a mind totally closed when it comes to self-advancement. Here three theories associated with physics have been highlighted as examples over which many have sold their souls but these are merely three examples out of a huge number across all branches of science. The status quo truly reigns and disturbing conventional wisdom is, to some, more reprehensible than high treason. However, where the public's hard earned money is concerned, members of that public should know just how flimsy some of the foundations of modern science truly are. Only then should these people be even asked to contemplate funding these very expensive toys with which some very highly privileged scientists can play!

17. What Is and What If.

Richard Lawrence Norman.

What is:

The happy fact remains: change is the only thing which life has to offer us. The hummingbirds hang suspended in space, floating, a magical excess of energetic expenditure, frozen amongst whipped air, waiting. A drink from the feeder, now rejuvenated, a dart of color speeds away into the sky, carving arcs of mad precision in a coordinated ballet of energy, sound and motion. The many varieties of caterpillar each become anew, their tube feet soon exchanged for wings, each blade of grass stretching toward the sun becomes a tasseled head of seed, and dies, the sun arching and sweet, soon rising to a boil at the apex of Summer's noon, pouring sheets of heat over the valley, then receding behind the distant hills to invite evening's cool, and the poetry is of one thing alone: becoming. The happy fact remains: change is the only thing which life has to offer us.

Process and transformation, are the basis of all things. Energy becomes mass, mass becomes energy, virtual particles appear and disappear, suns burn and die—the eternal is but the finite—ever changing.

In the world of science there are aspects which are otherwise—aspects which demonstrate a dynamic other than this singular bellwether of health itself: Change. Even so far back as the late 1800's the view into history was millennial, and the sight clear to interpret. From Bechamp's seminal work, *The Blood,* we read:

"An historian of the founders of modern astronomy recently related that the philosopher Cleanthus three millennia before our era, wished to prosecute Aristarchus for blasphemy, for having believed that the earth moved, and having dared to say that the sun was the immovable centre of the universe. Two thousand years later, human reason having remained stationary, the wish of Cleanthus was realized. Galileo was accused of blasphemy and impiety for having like Copernicus and following Aristarchus, maintained the same truth; a tribunal condemned his writings and forced him to a recantation which his conscience denied."

". . . I, Galileo, in the seventieth year of my age, on my knees before your Eminences, having before my eyes the holy gospels, which I touch with my own hands, I abjure, I curse, I detest, the error and heresy of the movement of the earth."

The case of Bechamp is cut of the same cloth. In Hume's *Bechamp or Pasteur* the grueling and tragic tale is laid out for all to see. Bechamp was a true genius with boundless energy, concerned with science alone. Pasteur was an animal of high ambition, although academically barely able to gain acceptance into the learned bodies which his abrasive personality would dominate. He ingratiated himself to the Emperor Napoleon, and became all but impossible to disagree with. However unassailable his personality, his science was lacking, if publically acclaimed. Over and over again the records demonstrate Bechamp's published work predates that of Pasteur's. In the cases of fermentation and silk worm disease the evidence is well past damning. However, history takes Pasteur's thefts and plagiarisms of Bechamp's work as lauded accomplishments, even as Pasteur's cruel indifference is utterly evident, a bully insensitive to thousands upon thousands of suffering animals given fake vaccines; without respect for the priority of another man's published work or anything else, save acclaim and money. In the end, Bechamp's deep insight and genius was left aside, and the pleomorphic aspects of disease with them (Hume, 2011). Reputation, personality, money and power make fodder of good science; fine men and their hard won work which could benefit many, are disgraced and suppressed.

The situation is little different today, and it appears that the scientific disciplines which are now as then funded, overseen and authorized by wealthy individuals and large powerful bodies, themselves composed of people filled with predictable human intentions and ambitions, have indeed created a situation which has to use the words of Bechamp, "remained stationary." In psychological terms, Science is ill, it is neurotic. Science demonstrates fixation. Let us take a brief accounting of some few of the current implications.

Monetary priority and medical practice: the 'patentable molecule.'

There are two sides to the conundrum of greed in medicine: the 'patentable molecule.' On one side, the drugs produced, just as Pasteur's lucrative yet deadly vaccines, carry with them a monetary incentive which affects bias toward confirmation of drug efficacy. That implies that drugs may be produced and sold which are ineffective and/or harmful so as to make money. The other side to this dirty coin, is the lack of incentive to bring forward treatment strategies or specific options which although effective and healthful, are not patentable and so, cannot extract money from the health of mankind.

Parkinson's and profit, un-patentable molecules and studies:

Parkinson's Disease (PD) is a common cause of neuro-degeneration in the geriatric population. This prolific and dread affliction may be ameliorated with a variety of substances which are unavailable for patent. This is not an assertion based in a soft-headed holistic naturopathic daydream. The following facts are extracted from detailed studies which are in the main available on the single most conservative source of modern mainstream orthodox science, the U.S. National Library of Medicine National Institutes of Health's archive at PubMed. Other sources below, are also from trustworthy peer reviewed journals. In place of the traditional reference list I will include a bibliography.

It should be noted that among the many compounds which are included below are some derived from cannabis, the international and local laws concerning which being quite arbitrary and various. In England doctors are legally and, I believe rightly, permitted to prescribe heroin in cases of severe pain, yet are not permitted to prescribe the much less dangerous drug cannabis, under any circumstance. One constituent in the highly complex assemblage of active compounds in cannabis, namely CBD, may well be efficacious in the amelioration of various pathologies from Parkinson's to seizure disorders, and causes no intoxicating side effects. It appears logical to reexamine the laws concerning cannabis and the rights of doctors to prescribe it, and/or its constituents. (The cannabis based pharmaceutical drug Sativex [GW pharmaceuticals] is the lone exception permitted for prescription in England to treat spasticity in multiple sclerosis). I do not recommend or advise any treatment strategy which does not adhere to the laws and legal codes where you reside.

Condensed facts [Cannabis/THC/CBD, Pregnenolone, Cinnamon, Thiamine, K2, D, Glutathione]:

Cannabis/THC/CBD and the uninvestigated role of pregnenolone:

a. From, *Modifications of neuroactive steroid levels in an experimental model of nigrostriatal degeneration: potential relevance to the pathophysiology of Parkinson's disease. Melcangi et al.*

"Among the neuroactive steroid levels assessed (i.e., pregnenolone, progesterone, dihydroprogesterone, tetrahydroprogesterone, isopregnanolone, testosterone, dihydrotestosterone, 3α-diol, dehydroepiandrosterone, 17α-estradiol, and 17β-estradiol), we observed a significant decrease of pregnenolone in the striatum."

b. From, *Cannabis (medical marijuana) treatment for motor and non-motor symptoms of Parkinson disease: an open-label observational study. Lotan et al.*

"RESULTS: Mean (SD) total score on the motor Unified Parkinson Disease Rating Scale score improved significantly from 33.1 (13.8) at baseline to 23.2 (10.5) after cannabis consumption (t = 5.9; P < 0.001). *Analysis of specific motor symptoms revealed significant improvement after treatment* in tremor (P < 0.001), rigidity (P = 0.004), and bradykinesia (P < 0.001). CONCLUSIONS: *There was also significant improvement of sleep and pain scores. No significant adverse effects of the drug were observed. The study suggests that cannabis might have a place in the therapeutic armamentarium of PD.* [Emphasis added].

c. From, *Pregnenolone Can Protect the Brain from Cannabis Intoxication. Vallee et al.*

"Pregnenolone is considered the inactive precursor of all steroid hormones, and its potential functional effects have been largely uninvestigated. The administration of the main active principle of *Cannabis sativa* (marijuana), Δ9-tetrahydrocannabinol (THC), substantially increases the synthesis of pregnenolone in the brain via activation of the type-1 cannabinoid (CB1) receptor." [Emphasis added].

d. There are antioxidant effects and others ascribed to CBD as well. From, *Prospects for cannabinoid therapies in basal ganglia disorders. Fernandez-Ruiz et al.*

"This CB(2) receptor up-regulation has been found in many neurodegenerative disorders including HD and PD, which supports the beneficial effects found for CB(2) receptor agonists in both disorders. In conclusion, the evidence reported so far supports that *those cannabinoids having antioxidant properties and/or capability to activate CB(2) receptors may represent promising therapeutic agents in HD and PD, thus deserving a prompt clinical evaluation."* [Emphasis added].

e. From, *Evaluation of the neuroprotective effect of cannabinoids in a rat model of Parkinson's disease: importance of antioxidant and cannabinoid receptor-independent properties. García-Arencibia et al.*

"In summary, our results indicate that those cannabinoids having antioxidant cannabinoid receptor-independent properties provide neuroprotection against the progressive degeneration of nigrostriatal dopaminergic neurons occurring in PD. In addition, the activation of CB2 (but not CB1) receptors, or other additional mechanisms, might also contribute to some extent to the potential of cannabinoids in this disease."

f. From, *Cannabinoids provide neuroprotection against 6-hydroxydopamine toxicity in vivo and in vitro: relevance to Parkinson's disease. Lastres-Becker et al.*

"In summary, our results support the view of a potential neuroprotective action of cannabinoids against the in vivo and in vitro toxicity of 6-hydroxydopamine, which might be relevant for PD. Our data indicated that these neuroprotective effects might be due, among others, to the antioxidant properties of certain plant-derived cannabinoids, or exerted through the capability of cannabinoid agonists to modulate glial function, or produced by a combination of both mechanisms."

We may conclude that Cannabis/THC/CBD may be helpful in the treatment of Parkinson's.

K2 and Mitochondrial function:

a. Parkinson's is a disease of energetic deficiency stemming from mitochondrial dysfunction. From, *PINK1 Loss-of-Function Mutations Affect Mitochondrial Complex I Activity via NdufA10 Ubiquinone Uncoupling. Morais et al.*

"A second hypothesis suggests that PINK1 has a direct effect on mitochondrial complex I, affecting the maintenance of the electron transport chain (ETC) resulting in decreased mitochondrial membrane potential and dysfunctional mitochondria."

And from *Mitochondrial Biology and Parkinson's Disease. Perier and Vila.* "Whether a primary or secondary event, mitochondrial dysfunction holds promise as a potential therapeutic target to halt the progression of dopaminergic neurodegeneration in PD."

b. Mitochondrial electron carrier, vitamin K2, rescues Parkinson's disease models based on this theory. From, *Vitamin K2 is a mitochondrial electron carrier that rescues pink1 deficiency. Vos et al.*

"We found that vitamin K(2) was necessary and sufficient to transfer electrons in Drosophila mitochondria. Heix mutants showed severe mitochondrial defects that were rescued by vitamin K(2), and, similar to ubiquinone, vitamin K(2) transferred electrons in Drosophila mitochondria, resulting in more efficient adenosine triphosphate (ATP) production. Thus, mitochondrial dysfunction was rescued by vitamin K(2) that serves as a mitochondrial electron carrier, helping to maintain normal ATP production."

———

We may conclude that K2 may be helpful in the treatment of Parkinson's.

Vitamin D:

Vitamin D has been demonstrated to slow the physical deterioration associated with Parkinson's. From, *Randomized double blind placebo controlled trial of vitamin D supplementation in Parkinson disease. Suzuki M, et al.*

"Compared with the placebo, vitamin D3 significantly prevented the deterioration of the HY stage in patients [difference between groups: $P = 0.005$; mean $\pm$ SD change within vitamin D3 group: $+0.02 \pm 0.62$ ($P = 0.79$); change within placebo group: $+0.33 \pm 0.70$ ($P = 0.0006$)]."

———

We may conclude that Vitamin D may be helpful in the treatment of Parkinson's.

Glutathione:

According to Dr. Julian Whitaker, from his newsletter of September, 2014:

"Glutathione is the major antioxidant produced in neurons and cells throughout the body. Oxidative stress and inflammation are implicated in the dysfunction and ultimate death of dopamine-producing cells. Restoring depleted glutathione stores slows this destructive process and improves symptoms in patients with Parkinson's. IV administrations helps ensure it gets into the brain.

I'll never forget one of the first patients we treated at the clinic with IV Glutathione. He had a significant tremor in his left arm and arrived in a wheelchair. After his second IV treatment, his tremor decreased and he was up and walking, albeit with an unsteady gait and his arms stiff at his sides. After his third infusion, he was walking more or less normally, with a confident stride, arms swinging—and no tremor."

Also see: *Reduced intravenous glutathione in the treatment of early Parkinson's disease. Sechi G, et al.*

"All patients improved significantly after GSH therapy, with a 42% decline in disability. Once GSH was stopped the therapeutic effect lasted for 2-4 months. 4. Our data indicate that in untreated PD patients GSH has symptomatic efficacy and possibly retards the progression of the disease."

Also see: *Glutathione and Parkinson's disease: is this the elephant in the room? Zeevalk et al.*

Nasal administration may also be effective. See: *Central nervous system uptake of intranasal glutathione in Parkinson's disease. Mischley et al.*

We may conclude that Glutathione may be helpful in the treatment of Parkinson's.

ilable

book

Thiamine:

From, *Long-Term Treatment with High-Dose Thiamine in Parkinson Disease: An Open-Label Pilot Study. Costantini et al.*

"CONCLUSIONS:
Administration of parenteral high-dose thiamine was effective in reversing PD motor and non-motor symptoms. The clinical improvement was stable over time in all the patients. From our clinical evidence, we hypothesize that a dysfunction of thiamine-dependent metabolic processes could cause selective neural damage in the centers typically affected by this disease and might be a fundamental molecular event provoking neurodegeneration. Thiamine could have both restorative and neuroprotective action in PD."

From, *High-dose thiamine as initial treatment for Parkinson's disease. Costantini et al.*

"Injection of high doses of thiamine was effective in reversing the symptoms, suggesting that the abnormalities in thiamine-dependent processes could be overcome by diffusion-mediated transport at supranormal thiamine concentrations."

From, *The Beneficial Role of Thiamine in Parkinson's Disease: Preliminary Report. Luong et al.*

"Five PD patients presented with stone face, right-hand tremors, Parkinsonian gait and bradykinesia with occasional freezing. Two patients presented with sialorrhea and the plasma transkelosase activity was low in one patient. All of the patients received 100 - 200 mg daily doses of parenteral thiamine. Within days of thiamine treatment, the patients had smiles on their faces, walked normally with longer steps, increased their arm swings, and experienced no tremors or sialorrhea."

_____

We may conclude that Thiamine may be helpful in the treatment of Parkinson's.

Cinnamon:

From, *Cinnamon treatment upregulates neuroprotective proteins Parkin and DJ-1 and protects dopaminergic neurons in a mouse model of Parkinson's disease. Khasnavis and Pahan.*

". . . However, oral treatment of MPTP-intoxicated mice with cinnamon powder and NaB reduced the expression of iNOS and protected Parkin/DJ-1 in the nigra. These findings paralleled dopaminergic neuronal protection, normalized striatal neurotransmitters, and improved motor functions by cinnamon in MPTP-intoxicated mice. *These results suggest that cinnamon may be beneficial for PD patients.*" *[Emphasis added]*.

_____

We may conclude that Cinnamon may be helpful in the treatment of Parkinson's.

Conclusion:

In the case of Parkinson's disease a safe, inexpensive, nontoxic, efficacious supplement might easily be developed based in this science. It may well offer substantial prophylactic protection against the onset of full blown symptomatology, and aid in the curtailment of disease processes when symptoms are evident. Those without active symptoms who have the dread LRRK2 mutation, or those with a family history of Parkinson's may be wise to take it, and those who display symptoms as well. Clearly, a diet rich in these pharmacologically active nontoxic compounds, may provide substantial benefit. I hypothesize as these studies are well known, that the only reason this obvious benefit has yet to be brought to fruition and these ideas are not in current clinical practice, is due to the fact that they are natural molecules and hence, cannot be patented. When money dictates medical practice, people remain ill and *pay*. Inexpensive effective treatments which do not benefit a large drug company or industry, are simply left to wither. This is why those effective treatments which are currently available are toxic and costly.

Oxytocin:

The category of 'unprofitable but safe' molecular constituents is large. I will choose very quickly oxytocin (OT) as an additional example. With antidepressant properties (Panksepp, 1998) and possible benefits extending from neurosis and sexual dysfunction to schizophrenia, alongside clear effects in creating neural plasticity, there are a great many who might benefit from different modes of treatment. I have constructed several such treatments but am unable to fund the studies to advance them. Why is this safe neuropeptide not already in clinical practice after years of detailed study?

"Although intranasal OT appears quite safe and tolerable, there are several practical barriers to its therapeutic drug development in humans. These include the lack of intellectual property ownership of the actual hormone, lack of US Food and Drug Administration (US FDA) approval for any psychiatric indication and challenges around the actual availability of the drug." [MacDonald and Feifel, 2012]—Oxytocin in schizophrenia: a review of evidence for its therapeutic effect.

The list of stated practical "clinical hurdles" articulated in that study is painfully weak. Only money has prevented this substance from serving the greater good and health of man.

Profit from poison:

The other face of the 'patentable molecule', this dirty coin of the realm in for-profit medical science, is to be found in toxic harmful compounds which although of little or no clinical use, do cause harm to those who take them and yield profit for the companies which develop and pedal them to consumers and physicians.

Statin drugs (such as Lipitor or Crestor), are not heart protective, they are a money making racket. They do lower cholesterol, but the benefits have been falsified. These drugs can CAUSE heart failure, and sabotage the energy production mechanisms of the cell. They cause the problems they are supposed to prevent. These deadly pills are, however, some of the very best selling drugs of all time.

An enzyme is blocked by statins which thereby suppresses the production of a coenzyme: CoQ10— that harms the ATP production process. The drugs are toxic to mitochondria. They interfere with K2 production. That leads to hardening of the arteries. These drugs can *cause* heart failure! Glutathione is interfered with leading to oxidative stress. As is known, statins are associated with cataracts, liver damage, kidney disease, cancer, sexual dysfunction, depression, memory loss, and diabetes. How have we citizens and many doctors been fooled?

"Relative Risk Reduction" statistical analysis has been falsely applied to create the impression that, what are ~one/two percent benefits…revealing a worthless treatment, which harms a great many, are "in fact" 30 and 50 percent gains in the amelioration of pathology. With annual lobbying for the pharmaceutical/health giants amounting to ~$235,107,261 in 2015, it appears, *the government is in bed with the corporations.* The modern system of money and scientific advancement is flawed, ugly and dangerous. An entirely new way to fund science is required.

From an important PubMed paper on the topic (Okuyama et al., 2015), we can see what this means for each of us:

"An impairment of selenoprotein biosynthesis may be a factor in congestive heart failure, reminiscent of the dilated cardiomyopathies seen with selenium deficiency. **Thus, the epidemic of heart failure and atherosclerosis that plagues the modern world may paradoxically be aggravated by the pervasive use of statin drugs**. We propose that current statin treatment guidelines be critically reevaluated." [Emphasis added]. [Statins stimulate atherosclerosis and heart failure: pharmacological mechanisms. Okuyama et al.]

Just in case you imagine that to be a fluke, a simple mistake from our benevolent and protective monetary-based authoritarian government and for-profit scientific and medical industries...please note the following: It is official that the top grossing drug in America (in 2014) was *an anti-psychotic*: Abilify. Complete with the usual anti-psychotic profile of side-effects, such as *permanent ticks and motor symptoms*: Tardive Dyskinesia. Now, prescribed for depression, typically with an SSRI (such as Prozac or Zoloft), which are themselves potentially associated with suicide upon withdrawal, and their own permanent condition, Tardive Dysphoria. Let's be clear: these "nonaddictive" SSRI drugs, do not themselves cause death upon withdrawal. SSRI drugs (used for depression and OCD) are only correlated with death via one of the most certain findings in all of psychiatry: low 5-HT is associated with suicide. Withdrawal therefore, may lead to death. Not an addictive drug. Simply know, if you stop from high doses, you may die by suicide. Taper very gradually, and only attempt withdrawal under a doctor's supervision, knowing, there may or may not be permanent damage. Now Abilify with its anti-psychotic profile of damage is also handed out like anti-psychotic candy for depression. American medicine...is a racket...nearly as lucrative as war. These drugs do most assuredly have a valid place in medicine, they are indispensable for those few who need them. Please do understand: using them as high dollar substitute jelly beans is not it. ~7 billion dollars in sales from Abilify, in one year (2014). Money makes for deadly, toxic medicine.

Where medicine intersects physics—new hope.

The greatest advance in our burgeoning understanding of the balanced biology which is health in the human animal, is to be found where physics intersects biological processes. Unfortunately, this pathway is fraught with danger. I will direct the reader to the chapter on Royal Rife for a taste of the reaction when a profound humanitarian crosses the sacred lines which separate the scientific disciplines, and dares to place the welfare of mankind before that of the potent powers which control what is, and what is not, acceptable scientific doctrine and truth. Rife understood that a knowledge of many disciplines is needed to accomplish any new and worthy thing, and of course he was correct. It is this which is most forbidden: to connect the threads of truth together and then create a new inexpensive way to benefit mankind which does not *first and foremost* profit the large corporations and governmental agencies and thereby support the entrenched paradigms which dictate the acceptable limits of science and the course of its efforts.

In the profound experiments of Nobel Laureate Luc Montagnier we see the essence of genius and hope, and also the essence of human intellectual cancer: the closed mind of science. The most deep wellspring of knowledge is to be found between the many scientific disciplines. Just as Rife demonstrated, a knowledge of many scientific disciplines is needed to gain headway toward the elusive goal of our deepest understanding. His was a mind not bound by petty greed and vanity, and he endeavored only to provide for mankind a safe, painless and inexpensive way to cure disease. His successful cure for cancer and many other diseases has of course been brutally suppressed and now mankind pays and suffers. A new approach which looks deeply, and in fact FINDS the answer, is that single result which is most tragically forbidden. To cure is forbidden...unless there is profit. The hope of mankind, has been bought and sold. I will show you where it resides, and how we may reclaim it.

Now Luc Montagnier has indeed found for us, a taste of the same: new insight. Of course, he has been denounced, shunned and insulted, his superb work discredited. He knew it would happen and even so, advanced along the correct pathway without hesitation. Like Rife, Montagnier is less concerned with the opinions and money offered up by others, and more so with the important work which will unriddle the deepest questions, and change the broken fate of mankind. Now the esteemed Dr. Montagnier, humanity's obvious benefactor, the discoverer of the AIDS virus and winner of a Nobel prize is himself feeling the ugly stain and sting of public scorn and professional rejection...because of one simple fact: he is *exactly right*. He has found a piece of essence, of deep and abiding truth, and so dear friend we may rejoice, for there is hope. I will tell you of it.

Physics understands what biology needs: electromagnetic fields and information.

All things...fields and particles alike, are based in information and 'observation.'

In my view, there is no conflict in these ideas and the moon is still there if you are not looking at it. Observation is simply informational exchange. Informational exchange is happening all the time through interactions throughout the physical system, and we as human observers are just a small contributor. No undue egoism or solipsism is needed to accept this truth: the universe is self-observing, and we, are part of the universe. The 'cognitive factor' is endemic to the system at all levels...information, is basic to physical processes.

Wheeler in 1990 stated: "It from bit symbolizes the idea that every item of the physical world has at bottom — a very deep bottom, in most instances — an immaterial source and explanation..."

As biology may be seen to take root in chemistry, and the basis of chemistry as Feynman was so eager to remind us may be found in physics, it is expected that biology also, must have information as its foundational basis. Indeed, it is so. Here, perhaps we have located the missing link in science, the connection between two disciplines, a nexus within which the essence of the problem may be caught unaware, and the simplicity found to unravel a great and tangled mystery.

Montagnier has demonstrated the informational aspects which sustain disease. It seems from my analysis that the cure for many diseases, from cancer to Alzheimer's and a great many more, may be found here. Just as in the case of Rife, the effect of this vital discovery was to isolate Montagnier from the funds and means he needs to advance, while he is heaped with scorn, ridicule and rude insult. Montagnier's revolutionary work is criticized on two counts:

1. It is said not to be repeatable in any other lab.

2. It is said to be a false result due to contamination.

Please note how similar this set of criticisms is to those leveled at Benveniste, a subject I will touch upon in a few paragraphs. In this case, Montagnier answered these criticisms in such a certain and clear way, as to leave the matter beyond dispute. He had invited an independent film crew from the media to record the experiment and watch each detail. He extracted the electromagnetic signature of a particular piece of DNA and sent that as binary information over the internet in excess of 1000 kilometres, then, had another *independent lab* in Italy receive the information and instantiate it into water memory via a simple electromagnetic process. Electromagnetic informational transfer is also the same way the bodily system works, in my understanding. The stunning result is clear and undeniable: he was exactly correct. The information once added to the test tube of pure water over 1000 kilometres distant, reproduced via water memory the exact encoding within DNA which was then synthesized via PCR, even though there was no template of DNA in the water! Information alone, once placed via an EM field into water memory created a piece of DNA and reproduced the encoding with an accuracy of 98 percent from raw PCR ingredients! Electromagnetic fields can be informationally encoded, and those fields affect aqueous systems, which receive the encoded information and interact with chemicals and biological structures to create the form specified. Biology is based in physics, and physics is based in information. The film crew's presence assures us there is no trickery, the second independent lab doing the PCR synthesis from water over 1000 kilometres distant, assures us of the experiment's verification at another facility, and most importantly precludes any possibility of contamination. Of course, the proof made no difference. Scientific orthodoxy simply turned up the insults. Now you may know with certainty: however well educated, those who discount Montagnier are shallow. The fact, has been clearly demonstrated, and the objections answered. He was right, the orthodox view is incorrect. Science is in the wrong. Science demonstrates something akin to a neurosis: fixation.

Clearly, the discovery of truths which offer clear promise of cure, or nontoxic treatments which do not imply drug sales and profits, such as this new science with its promise of diagnosis and the possible cure of many diseases with simple, noninvasive fields...is not wanted. However, although the massive scientific establishment will reap no benefit nor excise any undue profit from such groundbreaking work, the human profit, should it be developed, would be incalculable. The situation here is nearly akin to that of Rife. It could be different.

What If?

As I climb toward the noontime sky each step draws me higher, closer to the distant peaks. I pause, and look out over the valley within which my home is cradled. An amazing proliferation of motion and intricacy fill my eye, and life's enriching tapestry unfolds for me a vision of stunning clarity, each leaf and edge a painting etched in precise color wavering within a single wind, coherent and unified, yet, variant in the exact response of each leaf, and so, as a fractal relation in a multi-fractal system always tiny distortions added between the movements of one leaf to the next, the infinitesimal asymmetry of response creating a voluptuous effect, an effect as beauty is found in the asymmetrical distortions of classical Greek architecture, the errors are not errors, they are an essential intentional ingredient which creates beauty from the mundane, so was the breeze stroking the leaves of oak which dotted the distant hills, and I could see...all of it, from these many miles distant, now there within the sight, looking, watching...everything.

And there was more laid before me, hidden in plain sight and at the closest scales: Floating clear web tasting the breeze, the last drops of dew as round hearted prisms spattering the sun into giddy shards and then, a single leaf: within the intricate woven fabric of vein and fiber brocade, I could see the smallest structures and imagine the cells beneath, and so enter a labyrinth of detail and perfect intricacy, intimate and complex beyond measure—I am inside the maze of branched vein and green tissue, walking through intricacies of dendrite like webbing, and may look, and live, within the labyrinthian complexity and imagine the Minotaur awaits, a covetous aphid guards a drop of clear dew it has extracted from the vein of the world.

Oh how warm, intricately woven, changeable and subtle is life; health itself is a process, an evolution within the present toward the unknown. All of life is but change and growth, or we understand the fact of sickness, and decline. In its fixated state, science is revealed as Decadent. No less than that. What if it were different?

What if science had health and strength enough to look and admit, rather than refuse? This is the question which could liberate mankind.

We have recently published a paper: (Norman et al. 2016) *Quantum Information Medicine: Bit as It— The Future Direction of Medical Science: Antimicrobial and Other Potential Nontoxic Treatments,* [Richard Lawrence Norman, Jeremy Dunning-Davies, Jose Antonio Heredia-Rojas, Alberto Foletti]. Please recall the fact that Benveniste's work was brutally discredited as unrepeatable. We in our own way, have found otherwise. Here, you may see a *similar effect* in several highly replicable experiments which demonstrate that information associated with drugs may be encoded into water memory via a 7Hz carrier frequency and does indeed affect biological systems, much as the molecule from which the information was derived: *Bit as It*. Perhaps there is a new way to approach medical pharmacology without toxins. Perhaps information can be used instead of drugs to gain drug effects. What sort of effects have we found? Here is the abstract:

"Experimental evidence has accumulated to suggest that biologically efficacious informational effects can be derived mimicking active compounds solely through electromagnetic distribution upon aqueous systems affecting biological systems. Empirically rigorous demonstrations of antimicrobial agent associated electromagnetic informational inhibition of *MRSA, Entamoeba histolytica, Trichomonas vaginalis, Candida albicans* and a host of other important and various reported effects have been evidenced, such as the electro-informational transfer of retinoic acid influencing human neuroblastoma cells and stem teratocarcinoma cells. Cell proliferation and differentiation effects from informationally affected fields interactive with aqueous systems are measured via microscopy, statistical analysis, reverse transcription polymerase chain reaction and other techniques. Information associated with chemical compounds affects biological aqueous systems, sans direct systemic exposure to the source molecule. This is a quantum effect, based on the interactivity between electromagnetic fields, and aqueous ordered coherence domains. The encoding of aqueous systems and tissue by photonic transfer and instantiation of information rather than via direct exposure to potentially toxic drugs and physical substances holds clear promise of creating inexpensive non-toxic medical treatments".

Yes, effects are produced on malignant cells, neuroblastoma cells and stem teratocarcinoma cells, and even upon the stubborn and treatment resistant MRSA! As to the procedure being replicable, the reader may enjoy the following papers:

Antimicrobial Effect of Vancomycin Electro-Transferred Water against Methicillin-Resistant Staphylococcus aureus Variant. Heredia-Rojas et al.

Entamoeba histolytica and Tricho- monas vaginalis: Trophozoite Growth Inhibition by Metronidazole Electro-Transferred Water. Heredia-Rojas et al.

Antimicrobial Effect of Amphotericin B Electronically-Activated Water against Candida albicans. Heredia-Rojas et al.

Experimental Finding on the Electromagnetic Information Transfer of Specific Molecular Signals Mediated through Aqueous System on Two Human Cellular Models. Foletti et al.

Differentiation of Human LAN-5 Neuroblastoma Cells Induced by Extremely Low Frequency Electronically Transmitted Retinoic Acid. Foletti et al.

Yes, it appears Benveniste was discredited and ruined, although he was exactly correct. The implications are staggering. What hope lies hidden here beneath this error? Exactly what you might expect: nontoxic, inexpensive medical treatments which could help millions. All this may still be pursued and developed. To what end remains unknown:

(With modifications) From Norman et al. 2016:

"The following possibilities are just that: *possibilities.* The situation as it stands concerning our knowledge beyond the clear experimental evidence at present is plain: *We do not know.* Please review the following speculations with care, and assess the potential to be explored.

Unexplored Potential Benefits:

The future potential for inexpensive nontoxic drug-effect treatments, the possible alleviation of chronic pain and addiction are implied alongside delivery of the *effects* of drugs into the brain which themselves cannot cross the Blood-Brain Barrier (BBB). Future treatment strategies which currently remain undeveloped are therefore implied for diseases such as OCD and Parkinson's. Addiction of all sorts, from tobacco to heroin, may possibly be ameliorated. Drugs may potentially be subject to quantum replication yielding many doses from one dose of active substance. Those who are economically disadvantaged may, if this potential is realized, then have access to the effects of drugs which would not otherwise be available to them. *New approaches to antimicrobial therapies are implied.* Chronic pain, may potentially be addressed with information and so, perhaps without recourse to, or with less dependence on, addictive drugs. These potentials remain untested.

A Few General Points:

1) The BBB prevents many molecules from crossing into the system of the brain, so 5-HT cannot be delivered for OCD, and dopamine cannot be delivered in cases of Parkinson's. Many neuropeptides are also unavailable as vital therapeutic aids; 2) Addiction requires the administration of the very substance which creates the imbalance, be tapered in many doses to ease withdrawal, or, the pain of deep withdrawal results;

3) Protein folding is interactive with water structure. These techniques affect water structure. Each drug has a (structured) water signature. Long term research may well focus on defining the unknown relation between protein folding, water structure, electromagnetic distribution of quantum information, cancer and Alzheimer's.

These conditions/problems one and all *may* be amenable to this approach. Water easily gains access across the BBB, and/or a field may be directly applied. So, the entangled information associated with a drug or compound may be substituted for a drug, perhaps morphine, or dopamine. Now, as water (or a field) easily passes through or bypasses the BBB entirely, once encoded with the information and active effects of dopamine, a positive effect on Parkinson's is conceivably possible. Those neuropeptides which are currently undeliverable, with their subtle levels of behavioral specificity may now potentially also be available as therapeutic aids.

Perhaps, for chronic pain treatment and other such applications, a combined approach using the entangled information associated with a large dose in combination with a small dose of a real drug may be demonstrably effective. It is possible that addictive drugs may be avoided entirely. *Non-toxic informational drug effects may potentially help those afflicted with chronic pain.*

Quantum replication ("cloning") is implied: *A single dose of a drug may produce thousands of informational doses.* Drug costs could be reduced.

Potential for Addictive Amelioration:

1.   Addiction is created by the substitution of an external compound for an endogenous compound.

2.   Addiction's resultant self-sustaining homeostatic imbalance is reinforced with each additional usage of the drug.

3.   As a drug such as Methadone is addictive and the process of withdrawal without a substitute drug is a slow one, the treatment itself in both cases fosters the problem, and often fails. Imagine the number of people using nicotine patches.

4.   We propose that it may be possible to treat addiction in a new way which does not create the very problem it seeks to cure. The symptoms of withdrawal may well be quieted without a drug which creates more imbalance or the terrible pain of withdrawal, which leads to taking more drug to soften the blow or relapse. The addict may be administered water or a field infused with entangled information derived from their drug of choice. Their pain is thus reduced, and the problem not reinforced with more drug. This potential, now remains unavailable and untested.

Nobel Laureate Luc Montagnier recognizes the vital connectivity between quantum and biological processes. We believe he is correct.

Science has discovered many worthy and important things. There was a ~1.1 billion dollar cost for the 'discovery' of gravitational waves, as reported by *Scientific American.*

We submit to the reader, that an equally important and even more *practical human benefit* could come from detailed, stepwise, conservative experimentation to derive reliable replicable results in this new area: *Quantum Information Medicine.*

[Norman et al. 2016]

Now recall the new work of Montagnier. If science were to look here, what might happen? Imagine it. If this work were funded and closely investigated, we may soon have solved the riddle of the informational instruction set which creates DNA to sustain disease processes or health. That means two things:

1. A disease may be diagnosed in moments with a non-invasive scan.

2. A field may be applied to alter faulty encoding with correct patterning.

This is the *eventual potential.* Any disease which demonstrates resonance should be treatable and diagnosed in this way. As with the science of Rife, a resonant approach to disease and health is indicated. Here is found the common process basis of many diseases! There is a simple process nexus which may allow the informational alteration of fundamental disease dynamics without recourse to drugs, high priced treatments, or invasive techniques. In *Electromagnetic Signals Are Produced by Aqueous Nanostructures Derived from Bacterial DNA Sequences* he notes: "we have detected the same EMS in the plasma and in the DNA extracted from the plasma of patients suffering of Alzheimer, Parkinson disease, multiple Sclerosis and Rheumatoid Arthritis. . . . Moreover, EMS can be detected also from RNA viruses, such as HIV, influenza virus A, Hepatitis C Virus." [In this latter case after 20 nM filtration]. As I have stated, a great many diseases share the same mechanism of reproduction, and so may all be treatable and diagnosable in one simple way. Field effects, as Rife found long ago, may well hold the future of medical practice. Imagine a hand-held device which scans, finds resonant aspects of specific disease and after diagnosis, instantiates healthy patterning into the bodily system via an informationally encoded field, without the use of drugs. This is our future.

CRISPR technology alters DNA. The Chinese are applying it to human embryos. It seems natural to assume DARPA is using it toward no good end. The entire natural system is at its mercy, and gene drives have been constructed to force artificial genetic changes through entire populations of species. DNA is now a cut and paste affair. The unpredictable dangers of gene drives and CRISPR may find a better alternative here, by way of mimicking the means of natural informational transfer in the bodily system.

Also, if my analysis of the connection between epigenetic expression and pathogenic unconscious elements in the transference is valid, it may be possible to simply apply epigenetic information and treat mental illness! One might be able to discover the genetic instructions, the information to send which would allow genes to be expressed as chromatin, or converted to heterochromatin and shunted to the nuclear periphery where they may remain inactive. If so, a variety of conditions may be treated through information fields to safely ameliorate pathology, while leaving dangerous physical alterations of genetic material or harmful drugs aside. Perhaps, this is the future.

As Rife found a common resonant mechanism whereby he could treat and cure a great many diverse diseases, so has Montagnier uncovered a clue which will yield the ultimate prize should we be wise enough to look, rather than paint this deepest of all work with shallow scorn. *What if science would look?*

Let us approach the future with our heads held aloft, unaccepting of the broken situation, ask aloud and insist on a direct, thorough and honest answer.

1. Can information do the work of drugs? Can a computer network be established which will permit distribution of inexpensive and safe drug effects to all those in need, for little cost? Can parasitic greed be left out of medical care so the poor and rich alike may benefit? Information is all but free to replicate and distribute. Can we lift the wretched boot of greed from the health of the poor, and curtail the use of poisons where fields will suffice?

2. Can we create again, what Rife already had accomplished? Can we use frequency specific treatments to disrupt disease processes and cure cancer and other ailments in a cheap and painless way? Might we conduct science as Rife conducted science? Can we again find the pleomorphic processes which underlie a host of pathologies and

develop a proper cure as had been done so long ago, and brutally suppressed? To do otherwise, is clearly criminal. If you have technical skills or a lab which can work with filtered preparations, write me.

3. Now that Montagnier has successfully answered his critics, may we admit this and advance over the pathway he has cleared for us? May we look with great care to discover the multitude of diseases which can be diagnosed and cured by way of the physics of informational biology?

Conclusion:

Today, the various distinct scientific disciplines have each achieved within their own sphere, great and substantial progress. Now we are on the cusp of a profound revolution to be spawned through the unification of the entire of science, where distinct branches of study and truth will at last be understood for the intrarelated parts which they are. Physics provides a basis for chemistry, chemistry for biology and information provides, as Wheeler understood, a deep basis for physics. Indeed, the human animal in his state of disease and health is deeply akin to the most distant physical processes, all born of a common seed of energy and information. I assert: biology and the physical universe are *information pleomorphic*.

Encoded fields may one day replace toxic drugs and be used to restore balanced organization to the human bodily system. The union between physics, biology and information theory, if properly focused and practically applied, holds the next approach to humanitarian advancement and medical treatment. Might we raise our voices and ask of science a single question: *What if?*

Bibliography (suggested reading):

Bechamp, A. (2002) *The Blood and its Third Element.* Metropolis Ink. Metropolisink.com

Davenas, E., Beauvais, F., Amara, J., Oberbaum, M., Robinzon, B., Miadonnai, A., Tedeshi, A., Pomeranz, B., Fortner, P., et al. (1998) Human Basophil Degranulation Triggered by Very Dilute Antiserum against IgE. *Nature*, 333, 816-818.
http://www.ncbi.nlm.nih.gov/pubmed/2455231

Del Giudice, E., Tedeschi, A., Vitiello, G. and Voeikov, V. (2013) Coherent Structures in Liquid Water Close to Hydrophilic Surfaces. *Journal of Physics*: Conference Series, 442, 012028.

http://iopscience.iop.org/article/10.1088/1742-6596/442/1/012028

http://dx.doi.org/10.1088/1742-6596/442/1/012028

Diamond DM, Ravnskov U. (2015) How statistical deception created the appearance that statins are safe and effective in primary and secondary prevention of cardiovascular disease. *Expert Rev Clin Pharmacol.* doi: 10.1586/17512433.2015.1012494

http://www.ncbi.nlm.nih.gov/pubmed/25672965

Dunning-Davies, J. A discussion of structure and memory in water. *Hadronic Journal* **2012**, *35* 6: 661-669

Foletti, A.; Ledda, M.; D'Emilia, E.; Grimaldi, S.; Lisi, A. Experimental finding on the electromagnetic information transfer of specific molecular signals mediated through aqueous system on two human cellular models. *J. Altern. Complement. Med.* **2012**, *18*(3): 258-261. doi:10.1089/acm.2011.0104

http://online.liebertpub.com/doi/abs/10.1089/acm.2011.0 104?src=recsys&journalCode=acm

Foletti, A.; Lisi, A.; Ledda, M.; De Carlo, F.; Grimaldi, S. Cellular ELF signals as a possible tool in informative medicine. *Electromagn. Biol. Med.* **2009**, *28*(1): 71-79 DOI: 10.1080/15368370802708801

http://www.ncbi.nlm.nih.gov/pubmed/19337897

Foletti, A.; Ledda, M.; D'Emilia, E.; Grimaldi, S.; Lisi, A. Differentiation of Human LAN-5 Neuroblastoma Cells Induced by Extremely Low Frequency Electronically Transmitted Retinoic Acid. *J Altern Complement Med.* **2011**, *17*(8): 701-704. doi: 10.1089/acm.2010.0439

http://www.ncbi.nlm.nih.gov/pubmed/21721927

Foletti, A.; Grimaldi, S.; Lisi, A.; Ledda, M.; Liboff, A.R. Bioelectromagnetic medicine: The role of resonance signaling. *Electromagn Biol Med.* **2013**, *32*(4): 484-499. DOI: 10.3109/15368378.2012.743908

http://www.tandfonline.com/doi/abs/10.3109/15368378. 2012.743908

Foletti, A.; Ledda, M.; Piccirillo, S.; Grimaldi, S.; Lisi, A. Electromagnetic Information Delivery as a new tool in translational medicine. *Int J Clin Exp Med*. **2014**; *7*(9): 2550-2556.

http://www.ncbi.nlm.nih.gov/pmc/articles/PMC4211758/

Foletti, A.; Ledda, M.; Grimaldi, S.; D'Emilia, E.; Giuliani, L.; Liboff, A.; Lisi, A. The trail from quantum electro dynamics to informative medicine. *Electromagn Biol Med*. **2015**, *34*(2): 147–150. doi: 10.3109/15368378.2015.1036073.

http://www.ncbi.nlm.nih.gov/pubmed/26098527

Garcia-Segura, L. (2009) *Hormones and Brain Plasticity*. Cellular and Molecular Neuroendocrinology Laboratory, Cajal Institute, Oxford University Press, Oxford.
http://dx.doi.org/10.1093/acprof:oso/9780195326611.001.0001

Heredia-Rojas JA, Villarreal-Treviño L, Rodríguez-De la Fuente AP, et al. ANTIMICROBIAL EFFECT OF VANCOMYCIN ELECTRO-TRANSFERRED WATER AGAINST METHICILLIN-RESISTANT STAPHYLOCOCCUS AUREUS VARIANT. *Afr J Tradit Complement Altern Med*. 2015; 12(1):104-108

http://dx.doi.org/10.4314/ajtcam.v12i1.15

Heredia-Rojas JA, Torres-Flores AC, Rodríguez-De la Fuente A.O, et al. Entamoeba histolytica and Trichomonas vaginalis: Trophozoite growth inhibition by metronidazole electrotransferred water. *Exp. Parasitol*. 2011; 127: 80-83.

http://www.ncbi.nlm.nih.gov/pubmed/20603119

Heredia-Rojas JA, Gomez-Flores R, Rodríguez-De la Fuente AO, et al. Antimicrobial effect of amphotericin B electronically-activated water against Candida albicans. *Af. J. Microbiol. Res.* 2012; 6(15):3684-3689.

http://www.academicjournals.org/article/article1380805 042_Heredia-Rojas et al.pdf

Hume, E. (2011) *Bechamp or Pasteur*. Plasticine paperback, Australia. Plasticine.com

Leuner, B., Caponiti, J. and Gould, E. (2012) Oxytocin Stimulates Adult Neurogenesis Even under Conditions of Stress and Elevated Glucocorticoids. *Hippocampus*, 22, 861-868. http://dx.doi.org/10.1002/hipo.20947

Lin, Y., Huang, C. and Hsu, K. (2012) Oxytocin Promotes Long-Term Potentiation by Enhancing Epidermal Growth Factor Receptor-Mediated Local Translation of Protein Kinase Mζ. The Journal of Neuroscience, 32, 15476-15488.

http://www.jneurosci.org/content/32/44/15476.full

http://dx.doi.org/10.1523/JNEUROSCI.2429-12.2012

Lisi, A.; Ledda, M.; De Carlo, F.; Foletti, A.; Giuliani, L.; D'Emilia, E.; Grimaldi, S. Ion Cyclotron Resonance (ICR) transfers information to living systems: effects on human epithelial cell differentiation. *Electromagn Biol Med.* **2008**, *27*(3): 230-240. doi: 10.1080/15368370802269135

http://www.ncbi.nlm.nih.gov/pubmed/18821199

Lynes, B. (2004) *The Cancer Cure That Worked*. Marcus books. Queensville Ontario.

Monks, D., Lonstein, J. and Breedlove, M. (2003) Got Milk? Oxytocin Triggers Hippocampal Plasticity. *Nature Neuroscience*, 6, 327-328. http://dx.doi.org/10.1038/nn0403-327

Montagnier, L., Aissa, J., Del Giudice, E., Lavallee, C., Tedeschi, A. and Vitiello, G. (2011) DNA Waves and Water. *Journal of Physics*: Conference Series, 306, 012007. http://dx.doi.org/10.1088/1742-6596/306/1/012007

Norman, R. (2015*a*) (Semi)-Regressive Plastic Attachment Therapy. *Mind* Magazine. New Ideas section. http://www.mindmagazine.net/#!new-ideas/czpl

http://www.mindmagazine.net

Norman, R. L. (2016*a*) Homeostatic Conductance and Parasympathetic Basis Alteration: Two Alternative Approaches to Deep Brain Stimulation in Parkinson's, Obsessive Compulsive Disorder and Depression. *World Journal of Neuroscience*, 6, 52-61. http://dx.doi.org/10.4236/wjns.2016.61007

Norman R. L. (2016*b*) New therapeutic intervention and assessment tools: GSR, sexual dysfunction and the Peptide Assisted Therapy method—an applied therapy and mathematical metric of healing. *Mind magazine.* New ideas section: *http://www.mindmagazine.net*

Norman, R.L., Dunning-Davies, J., Heredia-Rojas, J.A. and Foletti, A. (2016) Quantum Information Medicine: Bit as It—The Future Direction of Medical Science: Antimicrobial and Other Potential Nontoxic Treatments. *World Journal of Neuroscience*, 6, 193-207. http://dx.doi.org/10.4236/wjns.2016.63024

Okuyama H, Langsjoen PH, Hamazaki T, Ogushi Y, Hama R, Kobayashi T, Uchino H. (2015) Statins stimulate atherosclerosis and heart failure: pharmacological mechanisms. *Expert Rev Clin Pharmacol.* doi: 10.1586/17512433.2015.1011125.

http://www.ncbi.nlm.nih.gov/pubmed/25655639

Panksepp, J. (1998) *Affective Neuroscience: The Foundations of Human and Animal Emotions.* Oxford Press, New York.

Note: A version of this article (and book) with active embedded reference links is available at:

https://www.researchgate.net/publication/30750800 1_Beyond_the_Veil?ev=prf_pub

18. Some Final Thoughts.

All the articles in this book have been concerned directly, or indirectly, with the influence of so-called 'conventional wisdom' on science. Specifically, the areas of science which have been addressed have been astrophysics and cosmology predominantly. The fact that all the research in these, as well as other areas is ultimately funded by a largely unknowing public has also been considered. If not actually stated, it has certainly been implied that this public has been kept uninformed about much that is going on in all these areas. The media, in the main, exercises discretion in mentioning alternative theories which could help understanding but might also damage the credibility and standing of many who openly uphold the status quo in science. Many might well wonder why this matters, apart from the fact that huge amounts of public money are at stake. However, any true scientist should be interested in finding the actual genuine solution to any problem considered. If this is the case then all should retain open minds and be only too willing to listen to the ideas of others. This is manifestly not the case in many spheres of activity. Admittedly, much of the discussion here has been concerned with theory and some observation but many of the observations which need to be made are extremely costly. Also, any experiments necessary to truly examine pieces of theory are extremely expensive to perform and so frequently may not be repeated. Firstly, this leaves huge question marks over many experimental/observational results. More importantly though, it means there is huge pressure on the scientists involved in these experiments and observations to obtain the expected results. After all, the notion of searching for these expected results is what helped secure the funding originally. Success could mean continued funding; failure could mean a cessation of funding. This is not to say that the scientists involved are not honest but, very often, interpretation of experimental and observational results is not a completely clear-cut affair and the possibility of subconscious erring on the side of personal caution cannot be completely excluded from an honest assessment of the situation. Also, in these days, when obtaining research funding can be more beneficial for a person's career than the prosecution of the actual research itself, it might be felt that even more pressure is being placed on individuals. It might be noted that this latter point is, nowadays, a very real issue and, in the interests of true science, it should be addressed by those in authority as a matter of extreme urgency.

As has been stated on many occasions, many might feel this does not matter too much in an area such as astrophysics/cosmology since the topic is far removed from the everyday lives of most people. However, if these problems do exist in areas such as astrophysics, they will be present, at least to some degree, in most, if not all, other areas of scientific research. Obviously this would have to include medicine and there people are dealing with the life and death of human beings, not the life and death of stars. Unfortunately, in medical research, some of the signs are not good. Recently, a top level medical researcher who had a fully established laboratory up and running apparently had a funding request refused because he had no room for any of the funding body's own research students. Even more worrying is that the final decision appeared to be taken by an administrator. Since the laboratory was well established and fully staffed, it is difficult to see how room could have been made for more research students. In any case, if the research is important, surely that is what should be being funded? Again, various stories concerning possible treatments for cancer abound but, when a leading cancer charity was asked if there was any truth in some of these, no answer was forthcoming, not even an acknowledgement of receipt of the letter requesting the information. Such behaviour does little to help the public perception of bodies such as this and ultimately will lead to a reduction in public financial support. Again, courtesy costs nothing but, possibly more importantly, if stories concerning supposed treatments for any medical condition are advanced, bodies that know whether, or not, such treatments are worthwhile have a duty to answer queries concerning them. If such treatments do work, there can be no harm in admitting such; if they do not work but could even be detrimental in some instances, then it is vital that the information is disseminated as quickly and widely as possible.

There was no response either to a query concerning the way research funding was distributed. Again this was a rather worrying point because of the reason for the query. This concerned continued funding for a researcher involved in apparently important work associated with the bladder but who had demonstrated a possible lack of knowledge of that organ in discussion with another worker in the field. The researcher concerned had challenged the validity of a diagram of the bladder produced by the other worker but the other worker's diagram was unquestionably correct. Hence, grave doubts about the original researcher arose and must exist still. Was there another reason for claiming the said diagram incorrect? If so, it would be both interesting and informative to hear it. The fact remains that this incident does raise further very real concerns about the way medical research is being funded.

Hence, from the outside, one is left with grave suspicions over the behaviour of research in medicine and possibly all other areas of science because of the actions of a few, specifically in astrophysics/cosmology. It may be felt by many to be totally wrong to condemn the many for the actions of a few but it should be remembered that, in many of the cases mentioned, and certainly in medicine, the research grants being mentioned are huge sums of money and, as stated on several occasions, this is ultimately being paid for by a largely unsuspecting, uninformed public.

It seems the time has come for there to be far more openness in science. Popular science books and popular TV science programmes should not simply push the views of conventional wisdom but should rather aim to give a more balanced view, acknowledging that alternative views and explanations do exist. All too often, people with alternative views are said to be cranks but this is, in reality, a very easy, convenient way of dismissing ideas which could prove troublesome to those in positions of authority, especially when the suspicion exists that one factor at play in obtaining that position is strict adherence to the dictates of 'conventional wisdom'.

Conclusion:

There is a veil of secrecy and control placed before the open mind. Its limiting effects can be observed to influence the entirety of human endeavor and achievement, from the abstract and elusive foundations of theoretical physics to the demonstrable and prolific harm incurred within medical practice, and further still, its pernicious effect reaches into the fundamental empathetic organizational basis of the human mind, and hence, also spreads hidden poison deep within the society which emerges from our collective efforts. The veil is everywhere, its influence is omnipresent. So ordinary has the error become, so loudly and with such assurance is the lie proclaimed, that it has vanished in plain sight. This cannot be permitted. We insist, to plainly observe the fact is itself to lay the first stone of the pathway toward the better answer. Money, power, ego, authority, avarice, 'reputation' and our resultant human history of control under the auspices of an immoral authority, have found for us a viewpoint we are to simply accept and believe without question. This falsehood has become a sort of herd mentality, an accepted 'conventional wisdom' meaning 'truth' borne out as a matter not of well-considered and hard won scientific fact, but instead, as a false creation of authority and consensus.

It is our hope that this brief work will begin to pierce the veil and allow the reader to consider the entire situation for themself. The future of man's knowledge, health, happiness and continuity stand clearly before us.   No more bright or perfect thing can be conceived, than what was once hidden.   A beacon has been lit, and the future of man beckons... from beyond the veil.

Biography of Authors:

Jeremy Dunning-Davies entered Liverpool University as a Derby Open Scholar in Mathematics in 1959, graduated in 1962 and obtained a Postgraduate Certificate in Education in 1963. He was awarded his Ph.D. for a thesis entitled, "*The Ideal Relativistic Quantum Gas*" in 1966, in which year he took up an appointment as an assistant lecturer in applied mathematics at the University of Hull. He became a full lecturer in 1968, and a senior lecturer in 1981. He moved as a senior lecturer to the physics department of Hull University in 2002.

In 1987 he met Bernard Lavenda from Camerino at a conference and a highly successful research collaboration ensued into the probabilistic foundations of thermodynamics. This led them on to investigate and question, some of the results associated with the thermodynamics of black holes. In 1996, at a conference in London, he met Ruggero Santilli from Florida and a new interest in new clean energies and the safe disposal of nuclear waste surfaced.

Jeremy Dunning-Davies has produced well in excess of two-hundred academic articles and his various books include: *Mathematical Methods for Mathematicians, Physical Scientists and Engineers* (1982); *Concise Thermodynamics* (First Edition) (1996); *Exploding a Myth* (2007) and *Neo-*

Newtonian Mechanics with Extension to Relativistic Velocities, co-authored with Dennis P. Allen Jr. Jeremy Dunning-Davies's interests have ranged from thermodynamics to the properties of the ideal relativistic quantum gasses which, in turn, led him to an interest in white dwarf stars and, eventually, astrophysics and cosmology. Also, he has been deeply involved with, and interested in, educational issues, something which is not too surprising for someone whose father was an eminently successful primary school headmaster in South Wales and whose mother and wife were/are both teachers.

Richard Lawrence Norman is the founder, and Editor in Chief of *Mind Magazine* and *The Black Watch: The Journal of Unconscious Psychology and Self-Psychoanalysis*. He is co-founder of Future Life Net, a group of highly advanced quantum physicists and artists who share a new human vision based on quantum physics. A writer, newspaper columnist and musician with degrees in philosophy and music, he is the author of books and scientific papers spanning philosophy, psychology, neuroscience, physics, verse and fiction. Richard Norman is a weekly contributor to The Ultranet's *BlogIQ*, a blog serving the gifted community, and a regular contributor to the Prometheus Society journal: *Gift of Fire*. Richard's papers, artistic contributions and scientific articles number well over two-hundred, and are featured in journals including *Quantum Matter*. His new psychoanalytic technique, *Native Psychoanalysis* may allow more rapid healing within a psychoanalytic model. Richard was a band leader for 20 years. He is a highly accomplished musician and an innovator in music theory. His *Time Travel and Other Illusions* CD redefines the use of the chromatic scale itself, and demonstrates a new approach to polyrhythm. His books include: *This New Day* [Philosophy/psychology]; *Mind Map* [Basic, brief new approach to Psychology]; *The Black Mirror* [psychology/philosophy]; *Ever Deeper Never Better* [Novel]; *Time Saw a Fly* [Novel]; *Enough* [Novel]; *The Tangible Self* [Advanced Psychology].

For a sample of Richard's scientific and creative work please enjoy the entire of *Mind Magazine* www.mindmagazine.net. The New Ideas section contains a video sample of his music and some of his papers detailing new medical treatments without toxicity, including quantitative approaches to the human unconscious, new methods proposed to treat degenerative nerve disease and a host of other contributions to the future of mankind. Scroll to the bottom of the New Ideas section page for the latest thinking.

Note: A version of this book with active links is available at:

https://www.researchgate.net/publication/307508 001_Beyond_the_Veil?ev=prf_pub

www.ingramcontent.com/pod-product-compliance
Lightning Source LLC
Chambersburg PA
CBHW071409180526
45170CB00001B/33